1977

FLOWCHARTING: Programming, Software Designing, and Computer Problem Solving

FLOWCHARTING: Programming, Software Designing, and Computer Problem Solving

BERNARD B. BYCER

John Wiley and Sons, Inc.
New York · London · Sydney · Toronto

Library of Congress Cataloging in Publication Data:

Bycer, Bernard B
    Flowcharting: programming, software designing, and com-
puter problem solving.

    Bibliography: p.
    1. Flow charts. 2. Electronic digital computers—
Programming. I. Title.

QA76.6.B9   001.6'423   75-16175
ISBN 0-471-12881-3
ISBN 0-471-12882-1 pbk.

Printed in the United States of America

10 9 8 7 6 5 4 3 2

This book is dedicated to my wife, Lil, who said, "No, never again." after my
first book.

PREFACE

This book is written as a text and a reference for persons already in the computer programming field or preparing to enter it. The book teaches flowcharting techniques and also instills in the reader an understanding of the power, rigor, elegance, and versatility of flowcharting as a discipline. By means of text and sample problems, this book teaches flowcharting as: 1.) an aid in logic and exposition; 2.) a tool in computer programming; and 3.) a necessity in the budding science of automated, computer-aided programming.

Five main uses will be found for this book:

1.) An undergraduate text in a one semester or two semester course in basic programming.

2.) A reference for graduate students and professional programmers.

3.) A reference for instructors in computer science courses.

4.) A reference for those engaged in program writing research or for those documenting the results of such research.

5.) A text for company sponsored continuing education courses.

The order of presentation in this text is quite straightforward. The reader is first indoctrinated in flowchart techniques and is then introduced to a wide variety of topics, each of which is discussed in terms of flowcharting. Chapter 1 begins by reviewing computer program problem solving and showing the basis and need for flowcharting.

Chapter 2 develops a dictionary of flowchart symbols using ANSI (American National Standards Institute) standards. Some history of the development of the standards is given so that the reader gains a perspective and an appreciation for the standardization process.

Chapter 3 covers flowcharting details and conventions. Flowchart simplification is treated, including the multi-level concept (system chart referencing more detailed composite charts).

Chapter 4 outlines the major hardware elements of a computer and explains serial and parallel operations. Computer speeds, scheduling times, and the potential for time conflicts and processing bottlenecks are treated using flowcharts.

Chapter 5 introduces the concepts of basic programming processes using five primitive flowchart symbols to represent three basic programming operations. A software module is defined to have three attributes: input, body, and output.

Program flowcharting is presented in Chapter 6. The material covered in this chapter is an extension of the previous chapter. Labelling and mathematical notation are used to identify data content and data locations, leading into counting and indexing. Looping is shown to be equivalent in several high level languages by flowcharting their respected DO, GO TO, and IF statements. They are considered equivalent, because their respective flowcharts or their logic flow paths are identical.

Problem solving by flowcharts is demonstrated in Chapter 7 by detailing work problem examples for computer programs and software system design. In the first half, high level language problems are flowcharted and run on large scale computer systems and minicomputer and microprocessor configurations. Complete computer programs are supplied for the reader to run at available data processing centers. In the second half, flowcharts are used to define system operation and specify software requirements.

Chapter 8 covers the current state and future prospects for flowcharting. Automatic program writing, flowcharting, and hardware design are covered.

It is assumed that undergraduate level readers are familiar with computer programming or are taking a basic course in programming. The first six or seven chapters will complement such a one semester basic programming course. A preferable arrangement would be to expand the one semester basic programming course into two semesters, so the instructor can integrate this flowchart material with his programming material and request that all homework problems be flowcharted for checking each student's coding. This two semester arrangement has been used by the author successfully. Class progress in programming is slower in the first semester than if two related subjects were taught. But speed is obtained in the second semester, because flowcharts are used routinely, thus expediting both the lecturing process and the homework. The overall result is that the class learns more programming, while each student becomes an experienced flowcharter.

Since a Curriculum-68* flowchart type course is not available, the text material complements the Curriculum-68 courses in the following manner. The material in Chapters 1 through 5 complements Introduction to Computing (B1) and Computer and Programming (B2) by equipping the student with the tools necessary for effective use of computers in the solution of problems. Chapters 6 through 8 provide the means for expanding the concepts and techniques of System Programming (I4) and flowcharting as a high level language for study in Programming Languages (I2). Modeling and System Simulation (A4) is one of the common applications for flowcharting and is used to a great extent for comparison of solutions to basic software/hardware design problems in Advance Computer Organizations (A2).

This book is helpful for students who would like to learn flowcharting without classroom instruction. The first five chapters are intentionally simple and can be used as a selfinstruction manual where the emphasis is on flowchart symbols per se. The reader need not have experience in programming languages. If used by professionals or graduate students, there is sufficient detail to support other Curriculum-68 courses and to assist in generating a professional software document whether it be a term paper, thesis, or a commercial software package.

BERNARD B. BYCER
Huntingdon Valley, Pennsylvania

*Communications of ACM, Vol. 11, pp. 151 to 197. Association for Computing Machinery, Inc., 1968.

ACKNOWLEDGEMENTS

The original manuscript in pencil copy was proofread and checked both for grammar and technical competence by Stephen T. Matrazek. The chore was exhausting and only a friend would stay with it to the very end. The first type-written copy, the one submitted to the publisher, was reviewed by Barry Longmire for technical excellence and Ray Dupell for organization, style, and quality standards. The final manuscript was edited and corrected by Joe Copestakes. Dr. Stuart A. Steele, Department Manager of Software Design and Development, made available the company's facilities in the following manner. The technical editing department under Don Higgs supervised the editing of the total manuscript. All the typing and art work was done by Charlotte M. Graham and Jane Silber. Eleanor K. Daly, the Chief Librarian supplied the more than 100 reference bibliography requests.

I received support from friends in the professional community and industrial firms and I would like to cite a few. Frank E. W. Ogle translated my German flowchart articles. Bull sessions with John Mehling, Paul Derickson, Tom Rabb, Joe Mark, and many others were enlightening and productive. Brad Stults and Jesse Grodnick of the Burrough's Wayne Facility supplied the ALGOL quadratic material on the Burroughs 6500. Dr. P. G. Anderson, Newark College of Engineering, supplied the PL/1 Sieve Problem on the IBM 360/65 system. James H. Davidson, Jr., RCA, furnished the COBOL preprinted form example that was run on the UNIVAC-Series-70 7055. Stephen A. Kallis, Jr., Digital Equipment Corp., supplied the program and flowcharts on FOCAL. My material on the Hewlett-Packard 2100 series was updated by John Stedman and Phil Gordon, Hewlett-Packard, Data Systems, for the HP21MX Computer Family presentation.

In obtaining support from industrial firms, the author wishes to acknowledge the publication permission granted to reproduce some of their copyrighted materials: Robert G. Whittmore, IBM; Richard J. Brady, Burroughs Corporation; Harry D. Wulforst, Sperry Univac; Stephen A. Kallis, Jr., Digital Equipment Corp.; and Ronnie Covington, Hewlett-Packard. Elsewhere in the book, permission to reproduce other copyright materials by authors, publishers, and technical societies are cited and referenced.

CONTENTS

1

2

3

FLOWCHARTING DETAILS 63
and CONVENTIONS

4

DATA PROCESSING 106

8

CURRENT STATE and FUTURE PROSPECTS

A

FLOWCHART PROGRAMMING SYSTEMS (Interactive)

B

ANALYSIS and SYNTHESIS of ALGORITHMS by FLOWCHARTS

1

INTRODUCTION

Flowcharting is being used for software and hardware expressions and for graphical communication in problem solving and presentation of information. A flowchart is an effective layout for planning, designing, and evaluating a variety of composite operations that are equally applicable to both hardware and software system design. Using a set of geometric symbols and usage conventions, it is possible to describe specific operations and procedures in a concise manner. The flowchart consists of an assembly of geometric symbols connected by lines representing information flow, and supported by annotations, mnemonic coding, tabulations, and identifications (Figure 1.1). Flowcharts are being used by novices as their particular means of communication and expression. Some of these have become classics, as the one purporting to describe the procedure of "getting up in the morning to go to work." These flowcharts may be novel and humorous, but they illustrate the explicitness and lucidity of the flowchart medium (Figure 1.2).

Flowcharts are relatively simple to learn and master, and they are useful and powerful tools. Actually, a flowchart can be precise and definitive, limited only by the ingenuity of the writer. Where verbal descriptions are subject to misinterpretation, the procedure or operation, as drawn on a flowchart, needs very little clarification. The flowchart can define specifications and expected results, leaving nothing to chance. There may be several alternatives at the detail level, but the system definition, operation, and performance are specifically delineated.

A flowchart is an effective means of communication, and it can compensate for limitations of written text by providing an effective display of meaningful relationships and straightforward flow direction. The area of

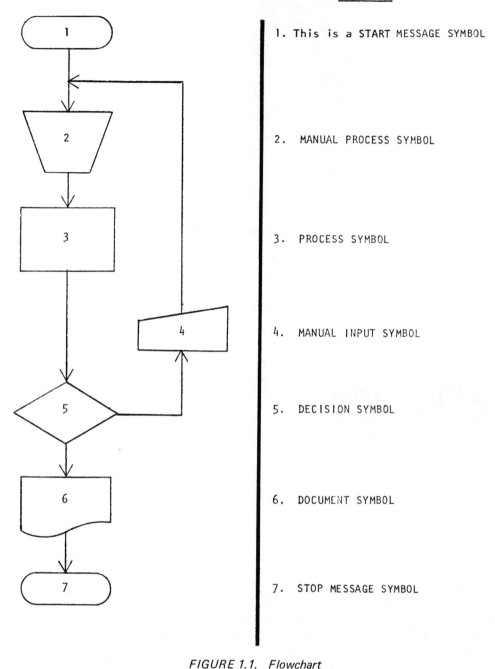

COMMENTS

1. This is a START MESSAGE SYMBOL

2. MANUAL PROCESS SYMBOL

3. PROCESS SYMBOL

4. MANUAL INPUT SYMBOL

5. DECISION SYMBOL

6. DOCUMENT SYMBOL

7. STOP MESSAGE SYMBOL

FIGURE 1.1. Flowchart

concentration is not limited to hardware and software in the computer field, but extends into many other areas. Flowcharting, as a dynamic language, is subject to evolutionary processes in the symbolic elements themselves and in their usage and application. Exposure of the flowcharting concept to a wide audience subjects the symbol elements to re-definition and new avenues of application from everchanging technologies. Flowcharts have a broad base of operation, permitting communication and documentation with a wide appeal and acceptance in numerous technologies. Flowcharting can simplify the communication problem between two technologies:

1. Software – Programming
2. Hardware – Equipment Design

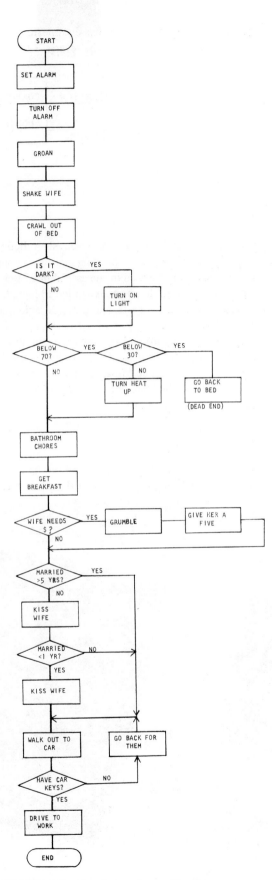

FIGURE 1.2.   Getting up in the Morning

3

The language barrier between these two technologies has disappeared. Flowcharts are mapped on a one-to-one basis into program statements[1] or logic operations[2] and the reverse is true as expressed below:

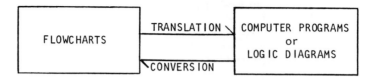

The flowchart language can be technology independent.

WHAT IS HARDWARE?

Hardware was present long before the word "software" was coined. A dictionary definition would suffice, but additional clarification is deemed advisable. Broadly speaking, hardware comprises all the electronic components, subassemblies, pieces, parts, and fasteners to be used in the assembling of electronic equipment (see Figure 1.3). As a vernacular expression, the term hardware has been expanded to include any and all devices, including metallic and nonmetallic, structurally rigid and semirigid parts, and fairly durable items. Computer equipment design groups are concerned with the details and construction of hardware, as well as communicating with programmers and users, in establishing the set of requirements and specifications for an operational computer system. Indeed, the hardware is the deliverable item. The concepts, analyses, studies, and other details of substantiation, feasibility, research, evaluation, and recommendations can be categorized as the backup paper (software) of a hardware design and implementation. Those things which are not hardware are consigned to the category of software. And as a corollary, that which is not software is hardware.

WHAT IS SOFTWARE?

"Software" is an interesting and provocative term that appeared on the scene concurrently with the digital computer. The general usage of "software" implies all things other than hardware (excluding people) required to

FIGURE 1.3. Hardware Logic Module

make a digital computer operational. The word "software" was immediately accepted in that jargon of the computer and processing personnel. It was readily associated with anything on paper. The phrase "Paperwork is our most important product" really sums it up and puts software on a pedestal. Software is a critical as well as expensive item that has given birth to a new industry — the software industry. Software requirements have grown so rapidly that software costs exceed the hardware investment in most installations. As long as the computer is in operation, it can be expected that additional funding for software will be required.

As a foundation for discussion, software refers herein to the programs and routines required to utilize and control the operation of the computer. This definition includes those operations necessary to prepare, translate, and convert data for input and output, and to control the computer in its performance of assigned tasks. Those routines which simplify, edit, and optimize storage requirements or running times are all categorized as software. Software also includes any program that is designed to solve a mathematical problem or a data processing operation, and any additional expediting and housekeeping details needed to accomplish these tasks. In terms of the deliverable end products, program listings (see Figure 1.4) and corresponding documentation (e.g., flowcharts) are the hardware of the software.

PROBLEM SOLVING

Problem solving is a complex task that requires many disciplines. In developing an insight on the subject, there are some common procedures which are followed whenever problem solving procedures are formulated. If the computer is used as a problem solving tool, the formulated problem solving procedures must be translated in a language for computer processing.

The term "algorithm" may be defined as a set of procedures for solving a problem. Flowcharts are a vehicle for representing such procedures. When the flowchart for a procedure has been developed, the logic inherent in it is then expressed in the form of a programming language that is the source input to the computer. The process of preparing the language equivalent of a flowchart or the conversion of the flowchart symbols into machine statements, directly or indirectly, is largely a mechanical one. The real challenge lies in the logic solution and not in the language or in the features of the programming language.

The average time spent by a programmer in using a computer as a problem solving tool is split equally between two major tasks:

1. Formulating the problem solving procedures to be executed by the computer.
2. Communicating the problem solving procedures to the computer.

Frequently, a little more time spent by the programmer in analyzing the problem and determining a solution will reduce the rework, patching, and modifying operations required to perform task two above. And the reverse is true. If little or no time is spent in the planning stage (problem analysis), complete chaos can result during task 2.

Problem solving includes five steps: problem analysis, flowcharting, coding, testing and debugging, and documentation. A brief description of the five steps for problem solving is given in the following paragraphs.

PROBLEM ANALYSIS

Contrary to popular belief, developing a software program is not accomplished by budgeting a major portion of the costs to the writing of a program, that is, filling in coding sheets. A balanced distribution between planning and implementing is suggested in Figure 1.5. The costs are split equally between the two. A further subdivision is given in average percentage values. The problem analysis phase includes defining the problem, gathering all the information and laying out a system solution. By reducing the system requirements to a flowchart, any external inputs become visible to the system analyst. Once the system specifications have been formulated (flowcharted), several solutions are possible, and research (tradeoff) studies can be conducted to assist in making a final selection (see Figure 1.6). The final selection is flowcharted and becomes the baseline system reference document.

The result of the analysis is a systematic descriptive solution covering three major areas as follows:

1. INPUT → 2. PROCESSING REQUIREMENTS → 3. OUTPUT

BURROUGHS B6700 ALGOL COMPILER, VERSION 2.6.000,   SATURDAY, 08/03/74, 06:45 PM.

```
                        Q U A D R A T I C
                        = = = = = = = = =

BEGIN COMMENT: THE SOLUTION OF THE QUADRATIC EQUATION                         1          B.0000 IS SEGMENT 0003

       A*X*X+B*X+C=0 WITH ARBITRARY                                                      0000:0000:0:0003
       REAL COEFFICIENTS A,B,C;                                                          0003:0000:0:0000
   REAL A,B,C,X1,X2,REALPART,IMAGINARYPART,                                              0003:0000:0:0000
   ZEROTOLERANCE,CONTRADICTION,DISCR,R;                                                  0003:0000:0:0000
   FILE PRINT(KIND=PRINTER);                                                             0003:0000:0:0000
                                                                                         0003:0005:0:LONG
   FILE CARD(KIND=READER);                                                    DATA       0003:0005:0:0000

   FORMAT INPUT(3F10.4));                                                     DATA IS 0005 LONG
   FORMAT OUTPUT(5F10.4,X5));                                                             0003:0000:0:0000
   ZEROTOLERANCE:=1.0@-8;                                                                 0003:0000:0:0023
   READ(CARD,INPUT,A,B,C);                                                                0003:0001:7:3
   DISCR:=B*B-4*A*C;                                                                      0003:0001:8:11
   IF ABS(A) GTR ZEROTOLERANCE THEN                                          2           0003:0001:9:11
   BEGIN                                                                                  0003:0001:A:2
     IF DISCR GTR 0 THEN                                                     3           0003:0001:A:25
     BEGIN R:=SQRT(DISCR);                                                                0003:0001:B:5
       IF ABS(B) GTR ZEROTOLERANCE THEN                                                   0003:0001:C:3
       BEGIN X1:=-(B+SIGN(B)*R)/(2*A);                                       4           0003:0001:D:1
             X2:=C/(A*X1)                                                                 0003:0001:D:15
       END B                                                                 4           0003:0002:1:3
       ELSE                                                                              0003:0002:1:4
       BEGIN                                                                             0003:0002:3:4
         X1:=R/(2*A);                                                                    0003:0002:4:10
         X2:=-X1;                                                                        0003:0002:4:4
       END                                                                              0003:0002:6:4
     END 2 REAL SOLUTIONS                                                    43          0003:0002:6:4
     ELSE                                                                    3           0003:0002:6:4
     BEGIN R:=SQRT(ABS(DISCR));                                                          0003:0002:7:1
       REALPART:=-B/(2*A);                                                              0003:0002:8:0
       IMAGINARYPART:=R/(2*A);                                                          0003:0002:9:0
     END 2 CONJUGATE COMPLEX SOLUTIONS                                       3C          0003:0002:C:5
   END A NOT EQUAL TO 0                                                     3C          0003:0002:C:15
   ELSE                                                                     2           0003:0002:D:2
   BEGIN IF ABS(B) GTR ZEROTOLERANCE THEN                                   3           0003:0002:E:3
     BEGIN                                                                  3           0003:0002:F:0
       X2:=-C/B                                                                         0003:0003:0:0
     END 1 REAL SOLUTION                                                    3           0003:0003:1:0
     ELSE                                                                               0003:0003:1:5
     BEGIN CONTRADICTION:=UNDETERMINED                                      3           0003:0003:1:5
     END SOLUTION UNDETERMINED                                              3           0003:0003:1:5
   END A EQUAL TO 0;                                                        2           0003:0003:1:5
   WRITE(PRINT,OUTPUT,X1,X2,REALPART,IMAGINARYPART,CONTRADICTION)
END.                                                                                    B.0000(0003) IS 0056 LONG
                                                                                        DATA IS 0005 LONG
```

FIGURE 1.4.   Program Listing

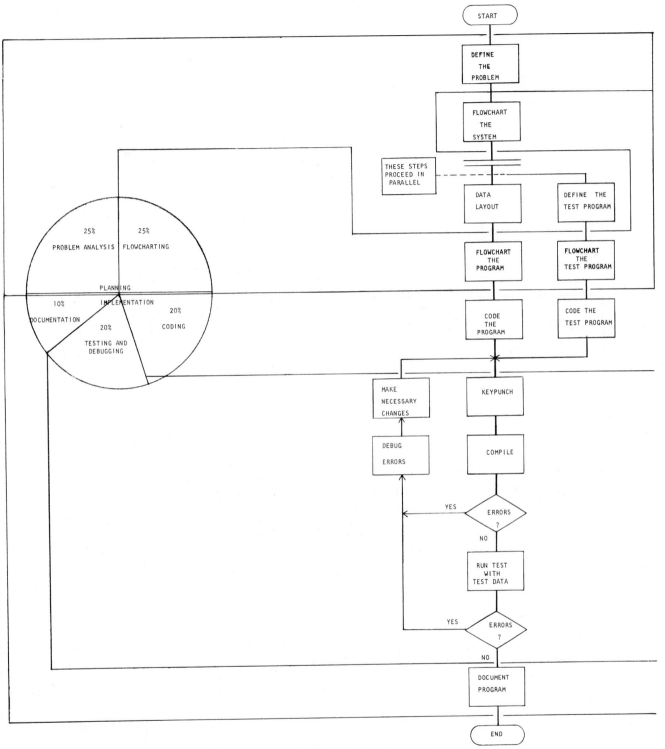

FIGURE 1.5. Software Problem Solving

The output is the desired results for a given input. The processing operations are the required algorithms to transform the inputs into the required outputs. In a payroll application, for example, the manner of presentation will include formats of the input and output records which are to be processed by the program, printer spacing charts which are used to illustrate a printed report, and some written narrative describing in detail the processing operations within the system and within each program included in the system.

SYSTEM REQUIREMENTS

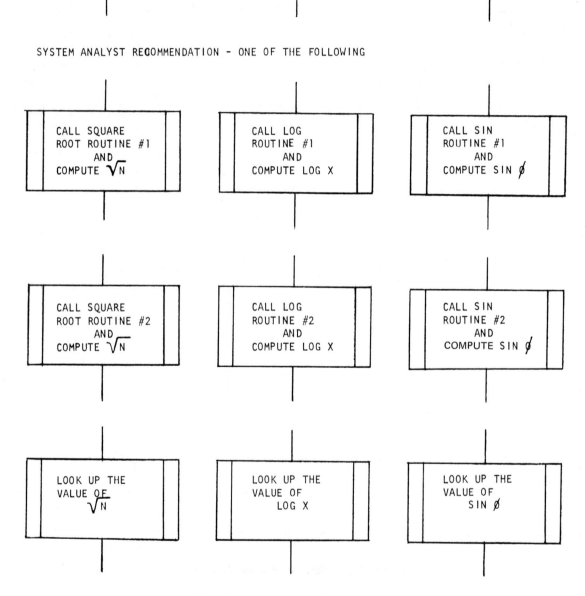

FIGURE 1.6. *Typical System Decisions*

A typical data layout for input and output is shown in this diagram.

Card Layout   INPUT DESCRIPTION

EMPLOYEE NUMBER (CHARACTERS 1-15)	EMPLOYEE NAME (16-45)	WEEKLY SALARY (46-55)	BLANK (56-80)

Printer Spacing Chart   OUTPUT DESCRIPTION

EMPLOYEE NAME (1-30)	BLANK (31-35)	WEEKLY SALARY (36-45)	BLANK (46-50)	EMPLOYEE NUMBER (51-65)	BLANK (66-120)

A more complicated configuration of data layout for input and output of a digital magnetic tape is shown in Figure 1.7. Of the numerous tape formats available, the system analyst must decide the content (parameters and their number length) of each record. The lower portion of Figure 1.7 shows the coding details of the variable record format. In examining the coding, the first field of the record shows number 180. This value is used to frame the record length (end points) for processing. The text material between the two record formats shows how the 180 is obtained. Two bytes are necessary to account for the value of 180.

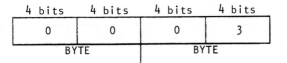

4 bits	4 bits	4 bits	4 bits
0	1	8	0
BYTE		BYTE	

Next 50 (two bytes long) denotes that 50 bytes of data will follow. Continuing, two more blocks are present that are 42 and 50 bytes long.

On the other hand another scheme is shown to process the same record length of data. The number three, defining the total blocks present, is used for indexing. To be consistent, two bytes are used to express this number

4 bits	4 bits	4 bits	4 bits
0	0	0	3
BYTE		BYTE	

The remaining details are repeated as shown in Figure 1.7.

As stated earlier, the total solution to problem solving for the problem analysis phase includes the specifications for the programmers to perform their tasks. A system flowchart is supplied to the programmer to indicate logic flow in the system (i.e., the sequence of events which occurs within the system to process the input data to obtain the required outputs).

Along with the system flowchart, the system analyst will normally provide the programmer with as complete a set of written instructions as possible with his recommendations for program writing. Once the set of specifications is received by the programmer, he must study in detail and understand every aspect of processing which can occur in the system. Normally, programs are written in response to controlled or noise free environment. Program constraints, parameter limits, and allocation of resources are all observed, but intermittent or transient errors are assumed to be nonexistent. Obviously, this is an invalid assumption. The programmer must understand possible error conditions and use recovery programs to maintain continuous computer operation.

It is mandatory that the programmer understand all of the processing operations before he attempts to lay out the logic of processing necessary to implement the system analyst's plans and recommendations. One of the difficulties encountered in data processing is that the programmer may not comprehend all of the requirements of the

FIXED RECORD TOTAL - FIXED RECORD LENGTH (F-F)

IBG	RECORD 100 BYTES	RECORD 100 BYTES	RECORD 100 BYTES	E B M	IBG	RECORD 100 BYTES	RECORD 100 BYTES	RECORD 100 BYTES	E B M	IBG

|←——— 300 BYTE BLOCK ———→| |←——— 300 BYTE BLOCK ———→|

FIXED RECORD TOTAL - VARIABLE RECORD LENGTH (F-V)

IBG	RECORD 150 BYTES	E R M	RECORD 47 BYTES	E R M	RECORD 100 BYTES	E R M	E B M	IBG	RECORD 98 BYTES	E R M	RECORD 99 BYTES	E R M	RECORD 100 BYTES	E R M	E B M	IBG

|←——— 300 BYTE BLOCK ———→| |←——— 300 BYTE BLOCK ———→|

VARIABLE RECORD TOTAL - FIXED RECORD LENGTH (V-F)

IBG	RECORD 100 BYTES	RECORD 100 BYTES	RECORD 100 BYTES	RECORD 100 BYTES	E B M	IBG	RECORD 100 BYTES	RECORD 100 BYTES	RECORD 100 BYTES	E B M	IBG

|←——— 400 BYTE BLOCK ———→| |←——— 300 BYTE BLOCK ———→|

VARIABLE RECORD TOTAL - VARIABLE RECORD LENGTH (V-V)

IBG	RECORD 100 BYTES	E R M	RECORD 90 BYTES	E R M	RECORD 80 BYTES	E R M	E B M	IBG	RECORD 100 BYTES	E R M	RECORD 200 BYTES	E R M	E B M	IBG

|←——— 273 BYTE BLOCK ———→| |←——— 302 BYTE BLOCK ———→|

IBG = INTER BLOCK GAP
EBM = END OF BLOCK MARKER
ERM = END OF RECORD MARKER (1 BYTE)

8 bits make a BYTE   n bytes make a WORD   n words make a RECORD   n records make a BLOCK

TAPE DATA RECORD AND BLOCK STRUCTURES

180	50	50 BYTES LONG	42	42 BYTES LONG	80	80 BYTES LONG

↑ BYTE TOTAL

2 BYTES	2 BYTES	50 BYTES	2 BYTES	42 BYTES	2 BYTES	80 BYTES

↓ RECORD TOTAL

3	50	50 BYTES	42	42 BYTES LONG	80	80 BYTES LONG

FIGURE 1.7. Variable Record Coding Details

program before he begins his program flowchart and coding. Sometimes when the program is being written, certain aspects of the program are not complete or not available. It is not always possible to anticipate all contingencies from the original design of the program. Making corrections to a program and adding routines when necessary is a very time consuming and difficult task. Experience shows that gross changes after the original program design will almost surely generate errors, because there is considerable patching or reworking of large program segments (e.g., salvaging of the initial program). Under these circumstances, astute advance planning and a good test program contribute to minimizing any need for serious rework.

FLOWCHARTING

The flowchart effort transcends both the problem solving tasks, planning, and implementing. Flowcharting is presented as a planning, designing, and an evaluating tool from the beginning until the completion of the software development program. As noted in Figure 1.5, the system analyst is responsible for developing the system flowchart that the programmer uses as his baseline reference. In turn the programmer generates his own flowchart for coding purposes. Another flowchart is generated to plan and code the test program to check out the computer program under test conditions.

When the programmer has determined that he understands all the processing requirements, the next step is to determine the logic to solve the problem by computer programming. The use of the flowcharting symbols and the logic to solve problems will be illustrated and explained in subsequent chapters.

CODING

If the program flowchart is quite detailed, the steps in the flowchart are easily converted into a computer program by the programmer. Commonly used programming languages include FORTRAN, COBOL, ALGOL, PL/1, ASSEMBLER LANGUAGE, and RPG. The programmer can use his program flowchart to check off his coding sheets to account for each logic operation required. Depending on his flowchart (shorthand) notation, the programmer may search up and down his coding sheets to ensure the flowchart logic is completely coded, (e.g., calls, subroutines, macros, etc.).

TESTING AND DEBUGGING

Testing and debugging costs usually match coding costs, as shown in Figure 1.5. Even in writing a small computer program or program module, testing and debugging are distinguishable from the coding process. Desk checking after a program has been compiled is a form of debugging. Everyone likes to have a 'clean' compilation. By doing this, the programmer can find errors both in the use of statements and in the logic that will be performed by the program. There are a number of manual techniques and computer listings to assist the programmer in debugging his program prior to execution. The tradeoff between machine time and programmer's time is too complex to be resolved here.

The preparation of test data is a difficult and ardous task. Sometimes a program cannot be adequately tested. A basic problem is always present: to choose the extent or degree of testing. The test problem is further complicated by requiring adequate performance specifications. Test and debugging will be amplified later, but for the moment, the reader should consider this fact. Testing (or system test) and debugging efforts require a separate written computer program and a programming effort from the computer program to be tested. Preferably, the persons who write the computer program should not write the test program. The test program flowchart and the program flowchart should be compatible and in agreement with the system flowchart generated by the system analyst. The three processes, coding, debugging, and testing, are vastly different and take place sequentially in time as shown in Figure 1.5.

DOCUMENTATION

Documentation is the process of recording the details concerning a computer program. The documentation will normally include the following: 1) a program flowchart; 2) a program narrative describing the routines (flowcharts optional) and programming techniques used in the program; 3) the source listing; 4) the formats of the data (input, output and internal files) which are processed by the program; and 5) a sample (load) run of the program including control cards. Part of the documentation is maintenance and diagnostic test routines.

Of all the tasks included in problem solving, documentation is frequently the most neglected. A thorough job of documentation is costly, time consuming, resisted by the programmer because he believes that it does not enhance nor contribute to the computer program itself. Anyone who has read a few program listings will agree that programmers are frequently very skimpy in the use of comment statements. Were it not for the department software standards, the comments by the programmer would suffice to assist him and be of little use to anyone else. Then, why consider documentation beyond the deliverable items: program listing; card decks; tapes; discs; user's manual; and a simple narrative?

It has been found that proper documentation of a program is absolutely vital to the smooth running of a data processing department. In many applications, changes must be made to a program once it has been put into the field and many times the programmer making the change is not the same person who originally wrote the program. Without sufficient documentation to indicate all the processing operation methods used in the program, it is quite difficult to make changes which will work properly. The maintenance programmer (program update and modification) installing a refinement or new facility in the program can seriously affect the reworked or modified program. It is incumbent upon the original programmer to supply sufficient documentation to ensure that any programmer assigned to make changes to the program can do so with a minimum of time spend on the job of learning how the program processes the data.

A good practice is to maintain a set of records throughout the development program under established software standards and conventions. All system, program and test flowcharts, and all program coding generated under software standards and conventions should eliminate the individualism of each programmer. The narratives generated for internal technical correspondence and design reviews using a standard outline are source material for documentation. Program documentation represents the total costs incurred in collecting and collating technical material generated during the development computer program, the staffing costs, and costs to supply printed or handwritten material along with the program listings.

SUMMARY

Flowcharting has been presented as a language having a set of geometric symbols for communication purposes as an alphabet. The next several chapters further detail a flowchart language. Flowcharts are used to record logic processes and to gain visibility from a total viewpoint. Flowcharts are a tool for problem solving and can be used for defining a problem, solving it, and evaluating the solution.

Problem solving through use of a computer includes five steps: problem analysis; flowcharting; coding; testing and debugging; and documentation. Planning (i.e., problem definition and solution) and implementing (i.e., designing, building, and testing) usually prove to be comparable in man hour costs.

Flowcharting should be initiated at the beginning and should be continued throughout the computer program development process. Basically, the system, program, and test flowcharts are working documents that can be included in the documentation package.* Software standards and conventions implemented at the beginning and continued throughout the program can reduce considerably the cost of preparing the documentation package.

REFERENCES

1. Michener, James Cope, "The Flowcharting Programming System," Ph.D. Thesis, Brown University, June 1970.
2. Bell, C. Gordon; Grason, John; and Newell, Allen, "Designing Computers and Digital Systems," Digital Press (Digital Equipment Corporation) 1972.

PROBLEMS

1. In the description of the problem analysis phase, the data layout comprised a punch card layout and printed sheet layout. What is a processing advantage using card inputs? What can be said about the printed employee listing?

   ANSWER

   A punched card would be prepared per employee and randomly inputted. The output listing would be alphabetically ordered and the employee's last name would be the keyword for searching.

2. Suppose there are two or more identical names including their initials. How would you determine the salary of the employee?

   ANSWER

   After finding the name, the employee number column would be used. Maybe the order of NAME, EMPLOYEE NUMBER, and then SALARY would be preferable.

*A system flowchart shows the flow of data through all parts of a data processing system. A program flowchart shows what takes place within a particular program in the system.

3. The layout for student records on one card is as follows:

NAME	ADDRESS	AGE	SEX	CLASS CODE	UNUSED

What codes and the number of columns might be used for SEX? For CLASS?

ANSWER

SEX   — one column — Code M and F or Numerics 1 and 2.
CLASS — one column — Code Alphanumerics

NUMERICS	ALPHABET	DESCRIPTOR
1	F	Freshman
2	S	Sophomore
3	J	Junior
4	S	Senior
5	G	Graduate or
or 5	M	MS
6	P	Ph.D.

By using numerics the ambiguous letter identification of Sophomore and Senior is eliminated.

In many cases the data layout, format, and coding have a great impact on data handling (i.e., packing and unpacking).

The next few questions concern Figure 1.7 Tape Data Layout for Input and Output.

4. Examine the figure and observe that the descriptors are called fixed and variable. Why?

ANSWER

The fixed word shows every word to be 100 bytes in length. The variable word varies from 80 to 200 bytes.

5. What is basic about the fixed word length and about the variable word length?

ANSWER

The fixed word length is a priori (known before) and therefore all formatting and coding details are built into the system — software and hardware.

The variable word length is not known in advance. Therefore, all formatting and data structure details must be included in the data and used to handle the data.

Notice the ERM (END of RECORD MARKER) separation between each variable word. This code terminates a preceding operation and is used to set up for the next operation (initialization).

In the fixed word length format, it is known in advance that each word length is 100 bytes. Therefore, it is not necessary to identify the beginning and end of each word.

2

FLOWCHART SYMBOLS

The flowchart is a method of expressing a set of procedures using what can be considered to be a procedural language. The symbols are the alphabet of the language; and with the associated text material and notations, these symbols constitute a vocabulary or dictionary. The sequence of flowchart symbols (procedures) is a composition of words that expresses an idea or thought algorithm. In the following pages, the ANSI (American National Standard Institute) flowchart symbol standard and industry flowchart nonstandard symbols are presented and defined in an expository manner.

Fortunately, the vocabulary and format structure of the flowchart language are much simpler to learn than a foreign language. The rules can be learned in the time it takes to study the ANSI flowchart standard. What is difficult is learning to use the language as an effective tool for expressing ideas. In the process of becoming familiar with the flowchart symbol set, the reader should pay attention to the very specific meaning and the generic description of the same symbol. The definition by examples is not restricted to software and hardware. The flowchart symbol vocabulary is presented as a universal language for logic expression similar to a mathematical language having its own set of standards, conventions, and symbolic notations. The symbols are defined and illustrated on a per-symbol basis. The rest of the book is devoted to use of the flowchart vocabulary for precisely describing logical procedures.

FLOWCHART STANDARDS

The major domestic standardization program dealing with computers and information began in earnest when a committee was formed in the 1960's. It was sponsored and administered by the Business Equipment Manufacturers Association (BEMA). The X3 standard program was developed by a committee composed of organizations representing three broad classes of interested parties as follows:

1. BEMA/Data Processing Group member companies (later called Data Processing Management Association — DPMA)
2. Consumers, trade associations, and government agencies
3. Specialized groups, usually professional and technical societies with special interests in computers and information processing.

During the formulation period, as many as 40 organizations comprised the three groups as members of the X3 committee which contributed to or were consulted in collecting, reviewing, and commenting on the available flowchart symbols. Initially the X3 committee drew up and proposed a set of flowchart standards that was published as American Standard X3 and reported by Robert J. Rossheim[1] in 1963. The working document issued by the committee covered the use of symbols in the preparation of flowcharts for information processing systems. It was authorized for publication primarily to elicit comment, criticism, and general reaction, with the understanding that such a working document was subject to change, modification, or withdrawal in part or in whole. Circulation was accomplished by journal and periodical publications.

Subsequently, in 1965 and again in 1966, 1968, and 1970, the X3.5 flowchart standard was revised and re-issued. The first three flowchart standard releases (1963, 1965, and 1966) were quite similar, having the same title (*Flowchart Symbols for Information Processing*) and their revisions were only minor. In 1963, the Input/Output symbol and the Manual Operation symbol were ⟍⟋ and ╱▢ respectively. In 1965, the Input/Output symbol and the Manual Operation symbol became ▱ and ⟍▢ respectively. Such areas as conventions, labeling, cross referencing, and repetitive symbol notation were not included until later revisions, including a new title (*Flowchart Symbols and Their Usage in Information Processing*). In 1965, the American Standards Association (ASA) was reconstituted as the United States of America Standards Institute and issued flowchart standard revisions in June 1966 and May 1968, while the ASA X3.5 Committee became known as the USASI X3.5 Standard Committee. Then in 1969, the organization name was changed to American National Standards Institute (ANSI) and this committee issued an updated flowchart standard version in September 1970. The 1968 flowchart standard revision was extensively modified; the title and organization name were changed, and changes were made in symbol identification, cross references, connector references, symbol striping, multiple paths, and multiple symbols.

The 1970 flowchart standard is an expansion of the X3.5–1968, which in itself is a revision of a 1966 standard on the subject. The symbol shapes contained in the X3.5–1968 are unchanged, but the 1970 standard defines 13 additional symbols. Moreover, several definitions and names have been modified slightly so that they conform more closely to standards approved by the International Standards Organization (ISO). One major addition to the 1970 flowchart standard concerns flowlines. Prior to 1970, crossing of flowlines was never permitted nor accepted under any circumstance. Now, crossing is valid. Based on present reports, the remaining conflicts between the ANSI and the ISO flowchart standards to be resolved are symbol identification, symbol and connector, and striped symbol cross referencing. This book will conform to ANSI standards.

The primary role of the X3.5 committee is to provide at the national level a vehicle for researching, surveying, developing, and promoting, flowchart standards. There is no legal force behind these standards. The basic underlying approval principle is consensus. In essence, those concerned with the scope and provisions of a proposed national standard ensure that the standard does indeed represent the thinking of the parties most affected by its provisions. Consequently, a standard is established having flexibility and versatility.

A discussion of the historical development of flowchart symbols for today's application and communication is purely academic. Nevertheless, knowledge of prior techniques, symbols, and conventions greatly enhances the flowchart communication, especially in areas that have not yet been defined or specified in an established set of standards. Normally current problems are identified through analysis in terms of prior knowledge. They are solved or resolved by the application of skill, creativity, and expression of an individual. The standard is intended merely as a guide to aid in its intended area and to prevent proliferation and provincialism, thereby permitting consensus rules to apply.

All the flowcharts presented in this book conform to the symbols and conventions and cross referencing in the ANSI X3.5 flowchart standard. However, there are cases presented that extend and enhance flowchart techniques. Their inclusion is necessary to develop the digital logic material and concurrency of operations, and in some cases to avoid the need for crossing of flowlines. The material presented in this book is generally in conformity with the Standard. Whenever the Standard is transgressed some explanation is given.

Acceptance of a nonenforceable standard whose track record has been five releases in less than 10 years is difficult. This is probably the reason that the Standard is still considered subject to change, especially in areas where symbols have gained acceptance in usage, but have not yet been incorporated in the present edition of the Standard. As an example, current flowchart templates include off-page-connector symbols, program modification symbols, transmitted tape symbols, and key operation symbols. Their usage has progressed to the stage of being included with the ANSI X3.5 symbols on a common plastic template. Actually, they are not in conflict, but await consideration and evaluation by the ANSI X3.5 committee.

Currently there are many nonconformities that may be classified in four categories:

1. Deviation
2. Augmentation
3. Objection
4. Violation

A deviation from the Standard usually arises from a definition that is either ambiguous or incomplete to some degree in the Standard. Obviously, there are different interpretations and if a particular interpretation is widely accepted, it becomes convention. By contrast, an augmentation or an addition to the Standard is an elaboration or extension of the Standard to cover a situation not provided for. Since the Standard cannot cover every point on the subject, those who use the Standard to the fullest extent may augment it with their own creations and contributions. Flowchart symbols form a living language, and more symbols are yet to be created or invented.

The present Standard permits several alternatives, and some may be ambiguous or may lead to misinterpretation. As an example is the ANSI basic symbol category, where not all the symbols listed under the Input/Output Symbols are considered to be truly I/O devices. Specifically the core symbol and the magnetic-disk symbol represent devices which are not universally considered as being in the I/O category even though the ANSI Standard so lists them.

Despite its shortcomings, the Standard is accepted nationwide, and deviations from the Standard should be avoided. Also flowline interaction and the double-arrow convention, even though allowed, should be avoided.

FLOWCHART SYMBOLS

Flowchart symbols are geometric shapes which describe an operation or process. The operation is conveyed by the shape (outline) of the figure, plus cryptic notations and annotations within the symbol. The simplicity of memory association and operation enhances flowchart appeal.

There are techniques for expressing an operation, or function, using a set of standard symbols promulgated by the X3.5 committee. In order to reduce any misunderstanding, the flowchart symbols presented here are consistent with those developed by the X3.5 committee on flowchart symbols for information processing. The flowcharting symbols have been grouped and classified here in a manner to facilitate ease of learning, usage, and retention by the reader, accelerating his comprehension and assisting in developing an applied skill to improve man-to-man communications. An attempt is made to complement existing standards by installing good judgment, and to develop good practices by supplying logical and justifiable rationale for a selection or a decision. Clarity, esthetic value, available aids (complete symbols or segments), physical symbol dimensions, legibility, supplementary knowledge, and orientation are vital and essential considerations in the preparation of flowcharts as well as the symbol selection, definition, and documentation.

BASIC SYMBOLS

The basic symbols are self-descriptive. They are fundamental elements without which even elementary flowcharts cannot be constructed. In the latest flowchart Standard (ANSI X3.5–1970) the flowchart symbols are grouped and classified in terms of functions or operations. The classification terminology is used to facilitate flowchart comprehension and application. The category of basic symbols permits flowcharting with a minimum of three symbols. The basic symbols are the foundation to which additional and special symbols have been added to complete the flowchart glossary. Furthermore, the symbol selection, definition and conventional usage permit expansion, modification, and removal if necessary, should the technology require it. The three basic elements are represented by the following geometric figures:

1. Line or Line Segments
2. Rectangle
3. Diamond Shape

FLOWLINE SYMBOL

The line or line segment is the means of connecting any two symbols to maintain the flow of information. Normally, the conventional method of reading, drawing schematics, and making column charts is from left to right. If the information exceeds one line or row, the process is repeated until a complete page has been scanned from top to bottom. Hence, the flowline symbol shown below represents the flowline functions for both directions.

When the flow direction is not left to right, or top to bottom, open arrowheads are placed on the reverse direction flowlines. Should increased clarity be required, open arrowheads can be placed on normal direct flowlines.

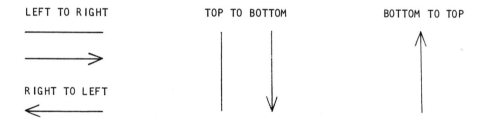

When flow is bidirectional, the approach taken here, and detailed later, depicts that each appropriate symbol (excluding flowlines, annotation symbol, communication link, etc.) has an entrance (input) and an exit (output). The single line or double line with open arrowheads used to indicate both normal direction flow and reverse direction flow should be avoided. In the former case, one arrow missing of a double arrow single flowline is a valid and acceptable notation. The input/output approach delineated later eliminates this typographical error. The double line should be reserved for parallel or simultaneous operations. The double line is a carryover from the bus line (multiple paths) from hardware design techniques. The double line bidirectional flow should be avoided for the same reason as the single line bidirectional flow.

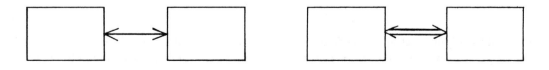

Present standards (ANSI X3.5–1970) permit flowline crossings provided no logical interrelation is implied. (See next page.)

Heretofore, this convention was invalid and in conflict with previous flowchart conventions. The standard flowchart extension requires further description and definition in areas of flowline functions and they are provided here. In cases where two or more flowlines join, provision is made to accommodate the following conditions:

1. Two or more incoming flowlines and one outgoing flowline.
2. One incoming flowline and two or more outgoing flowlines (not covered in the Standard).
3. Multiple incoming and outgoinging flowlines at a junction. (These shall have arrowheads near the junction point for flow indication.)

For item 1 above, the Standard is given and its corresponding interpretations are shown as follows:

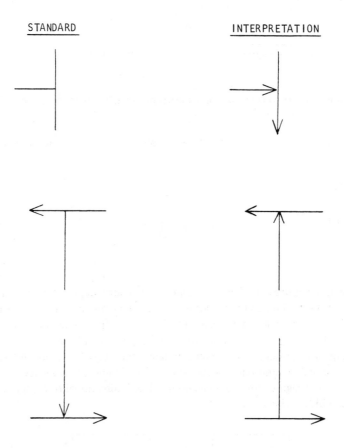

Item 2 is not specifically singled out and defined by the Standard. Instead multiple logic paths (ANSI X3.5–1970) and the parallel mode (ANSI 3.5X–1970) symbols do indicate a possible configuration. Also, the one incoming flowline and two or more outgoing flowlines are functionally switching or branching operations, provided they are sequential.

EXTENSIONS OF THE STANDARD   INTERPRETATION

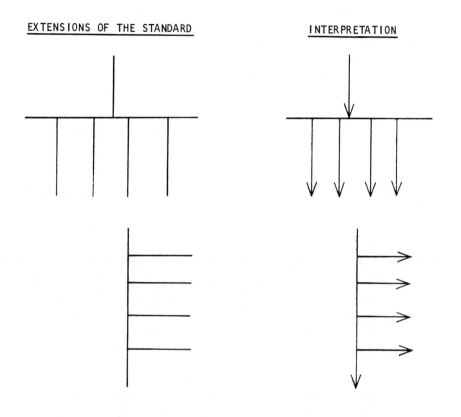

This symbol configuration has been used for parallel operation when the details associated with each line are defined. Generally speaking, concurrency may be inferred by examination. In many cases, especially in real time systems, concurrency is present and the above symbol is used both for sequential and simultaneous operations. Typically, one operation is initiated and that branch is selected. A code may be modified and the selection is repeated with another code while the previous selection is still in operation. Sequentially, each flowline is switched adding a new operation while retaining one or more of the previous operations. Under these conditions, concurrency is present.

A typical illustration for item 3 is diagrammed:

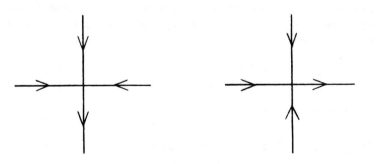

As noted above, both examples have three inputs and one output. A preferred configuration and more elegant style are shown:

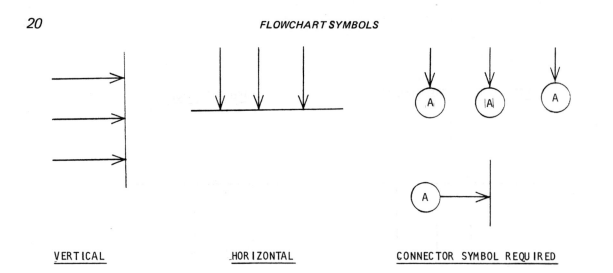

VERTICAL   HORIZONTAL   CONNECTOR SYMBOL REQUIRED

The connector symbol (shown above) has not been defined as yet, but its use can be discerned. Note that it eliminates junction points and flowline crossings. Here, and in the cases to follow, the flowchart symbols are found to be versatile enough to allow presenting an operation (function or process) in a number of ways. In keeping with this statement, item 2 could be implemented by using the branching symbol to define all of the conditions for multiple outputs. In any event, if the flowline symbol fails to convey the user's message or is subject to misinterpretation, it is improperly used.

PROCESS SYMBOL

The rectangular box is used to define an operation, a process, a function, or a combination (or group) of appropriate statements having a singular identity, or heading. It is a means of encircling or enclosing an idea(s) using a straight edge rather than attempting to draw a circle freehand. The rectangle can depict all the intended operations required for flowcharting. It is used to describe all operations for which special symbols are not designated.

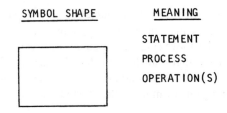

SYMBOL SHAPE   MEANING

STATEMENT
PROCESS
OPERATION(S)

The process symbol can represent any kind of processing operation or group of operations. This would include any algebraic and mathematical expression. The mathematical notation of equality such as $A = B$ has its counterpart as $A \leftarrow B$ in an assignment expression.

The expression appearing on the right of the arrow may be a single variable, a simple constant, or a complicated expression.

The letters A and B are labels or names that are used to represent A or B and each is stored in a memory device (or cell).

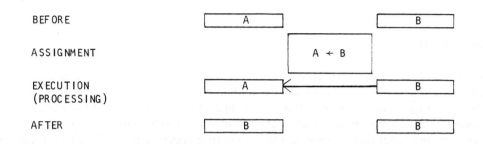

The location of A or B is the address of the memory cell. The arrow direction means that the expression on the right side now occupies or replaces the contents on the left side.

The addresses for A and B remain as they were before the operation. However, if you visit the address of former A, you will find the value of B located there. Also, the address of B will still retain the value of B. Thus A has been made equal to B, because everytime you address A you get B.

The addition of two numbers to obtain a sum will reinforce this assignment notation.

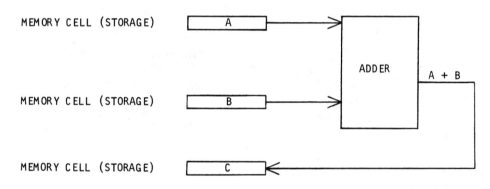

Here three memory cells are used to store three numbers, A, B, and C. The sum of A and B is located in C. Now two numbers are added using two symbols, and two storage locations.

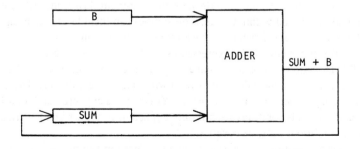

The logical configuration shown above is more efficient in that it accomplishes the same logical operation with fewer components. The initial value is stored in the SUM designated register and its contents added to the contents in the B register and stored in the SUM register. This operation can be expressed as

$$SUM = SUM + B$$

Mathematically, the equality is only true if B is zero. This conflict or possible misinterpretation can be avoided by use of the following convention:

$$SUM \leftarrow SUM + B$$

Contents	Contents of SUM
of SUM	and B before
after the	performing the
operation;	operation;
time — now	time — before

DECISION SYMBOL

The diamond shape (or rhomboid) is probably the only symbol which serves to truly distinguish a flowchart from a block diagram of other means of graphic presentation. What is implied is simply this: the very existence of a diamond symbol on a logic diagram or chart signifies it to be a flowchart. This symbol alone characterizes a flowchart drawing even when other geometric symbols are present in the drawing. Generally speaking, any of the flowchart functional or operational symbols need only two connecting lines (one for input and one for output). However, the diamond shape symbol needs at least three: one input and a minimum of two outputs. Herein lies the uniqueness of the diamond shape symbol hereafter called the Decision Symbol. This symbol shown below represents a decision or switching type operation that determines which of two (or more alternate paths) are to be followed.

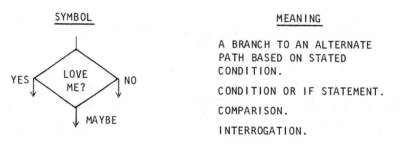

SYMBOL

MEANING

A BRANCH TO AN ALTERNATE PATH BASED ON STATED CONDITION.

CONDITION OR IF STATEMENT.

COMPARISON.

INTERROGATION.

It then follows that if two exits are present, then one of them is by definition not the normal direction, but a deviation or change of direction.

Anytime a deviation of direction (not multiple paths) from a sequential or operational series is depicted, the decision symbol is used to branch or switch direction to one of two or more alternate paths possible. This decision is based on a comparison, a test of a tag or flag, a balance test or sign indication, a program or operation switch, an index or jump instruction (computer), or an equipment insertion, inclusion or exclusion as the case may be.

The notation used within the decision symbol can include the equal sign (=) and a number of mathematical relationships. The equal sign can have a two output or a three output configuration, depending on the expression. The A = B test can equally be expressed as A : B. The A : B is read as "A is compared to B" or "Compare A with B." The interrogation and comparison test are shown in Figure 2.1 along with mathematical notations expressing specific logical decisions over a wide range of conditions. The decision symbol is an interrogation box. It asks a question and a reply of at least a Yes or No is expected and identified. Where a simple answer is not sufficient, multiple solutions are not only expected, but must be identified with the utmost of clarity, and all possible conditions must be covered.

A series of questions and answers using the decision symbol is shown in Figure 2.2. Any student in the computer science curriculum knows that the whole "Sign On" procedure at a remote computer terminal installation takes less time than it does to read the flowchart in Figure 2.2, provided everything is in working order. The same student knows what a horrendous problem it is to do a homework assignment when things are not just right. In the latter case, it can take several hours and be very exhausting. The left most column in Figure 2.2 shows the normal "Sign On" procedure at a remote terminal. All the flowchart columns to the right of this column depict possible problems. Sometimes the telephone line can be busy. Sometimes the terminal equipment can be

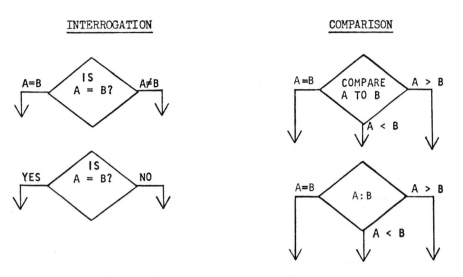

FIGURE 2.1. *Decision Symbol Example*

malfunctioning or the computer center equipment may not be working correctly. In attempting to isolate an equipment problem, the user may enlist the computer operator at the computer center to help solve or locate the problem. There are times when the computer center is not available to the student even if he attempts to use a high priority code. The computer center can be down for repairs, maintenance, or modification.

The decision symbol by name requires a decision to be made without equivocation. Therefore, all prior steps taken shall enable definition of the output domain (total outputs) completely. Any lack of completeness or rigor weakens the set of procedures presented or the logical path taken. In a two output decision symbol configuration, a single category is assigned to one output and the remaining items are all grouped together and assigned to the other output. In this manner the output domain is totally defined. To obtain more outputs and still have control of the output domain, a tree of decision symbols is constructed in the following manner.

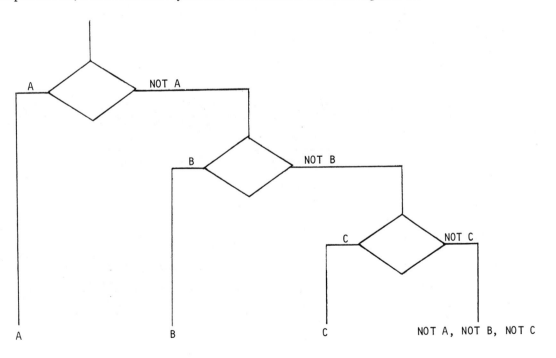

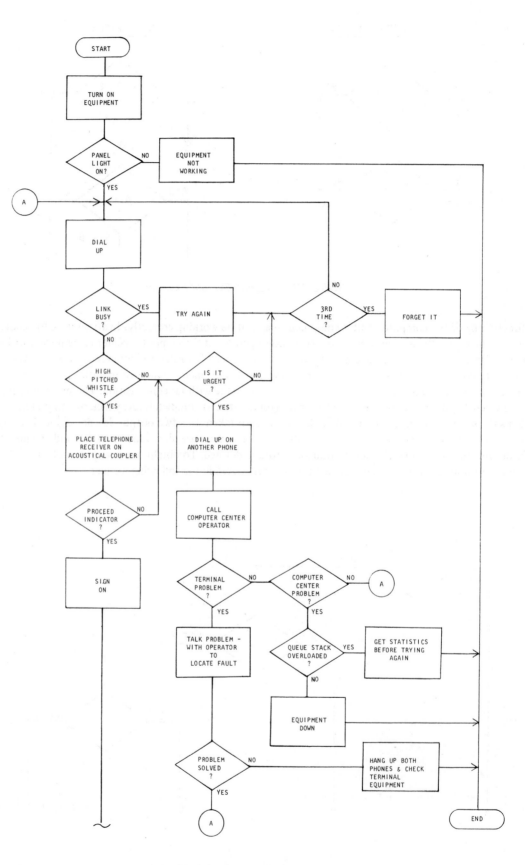

FIGURE 2.2. "SIGN ON" Procedure

ADDITIONAL SYMBOLS

As we have seen, the basic symbols are the line, the rectangle, and the diamond. The additional symbols are in reality extensions of the basic symbols and for clarification purposes. They serve to augment, supplement, and unify existing annotation techniques.

CONNECTOR SYMBOL

The connector symbol allows the maker of a flowchart to break a flowline while maintaining flowchart continuity. The symbol is used in place of flowlines which would cross other flowlines (Figure 2.3), and it can simplify an otherwise complex diagram. A circle is used as a connector to identify continuity of a single line(s) or data flow.

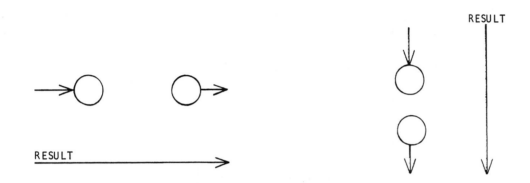

Obviously, a minimum set of two connectors is required when a flowline is broken. It follows that proper identification is necessary to identify the two terminals of flowlines for continuity. The connector circle definition is expanded to include the representation of a junction of several flowlines with one flowline or the junction of one flowline with one of several alternate flowlines by a set of two or more connectors. Where necessary, for clarity, it may be desirable to avoid having more than one line entering a symbol (or box) by using a connector. The connector symbol is a means of entry to or an exit from another page, or even another document (see Figure 2.4). It is used to connect different levels of flowcharts when there is a hierarchy. These details and other flowchart conventions are covered in the next chapter.

ANNOTATION SYMBOL

The annotation symbol allows the maker of a flowchart to add parenthetical remarks, to elaborate on statements contained in boxes or enclosures, or to insert notes to aid the reader.

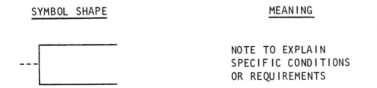

The horizontal broken line attached to the vertical solid line may be drawn either to the left as shown, or to the right when the symbol is rotated 180 degrees about the vertical axis. It is connected to any symbol at a point where the annotation applies. The basic purpose of a flowchart is to simplify the understanding of an operation, program, or system. If a short note or reference can add clarity to a flowchart, this symbol can be used (see Figure 2.5). However, extensive use of this symbol should not be necessary. Repeated use of the annotation (e. g., approximately one-to-one ratio with logical operation flowchart symbols) implies either insufficient information or a flowchart that is too brief for its audience. The writer of a flowchart should use more flowchart symbols to detail the logical processes and be a little less dependent on the annotation symbol. The annotation symbol should

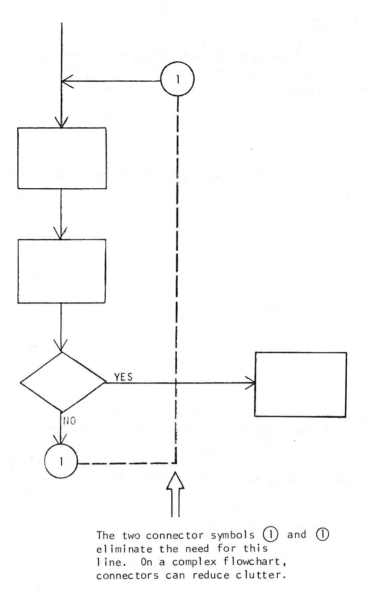

The two connector symbols ① and ①
eliminate the need for this
line. On a complex flowchart,
connectors can reduce clutter.

FIGURE 2.3. Line Crossing Elimination

be reserved for tagging, flagging, or highlighting an important point rather than cluttering up an otherwise adequate and presentable flowchart.

TERMINAL SYMBOL

There are a number of start/stop operations that are significant enough to warrant a special symbol to denote this function. The oval symbol shown below represents a terminal point in a system or communication network at which point an operation ceases to be active and dynamic.

SYMBOL SHAPE	MEANING
⬭	BEGINNING, END, OR POINT OF INTERRUPTION

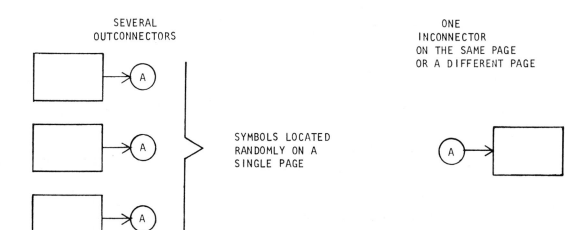

SEVERAL
OUTCONNECTORS

ONE
INCONNECTOR
ON THE SAME PAGE
OR A DIFFERENT PAGE

SYMBOLS LOCATED
RANDOMLY ON A
SINGLE PAGE

THE ABOVE CONNECTORS ACHIEVE THE FOLLOWING FLOW CONTINUITY

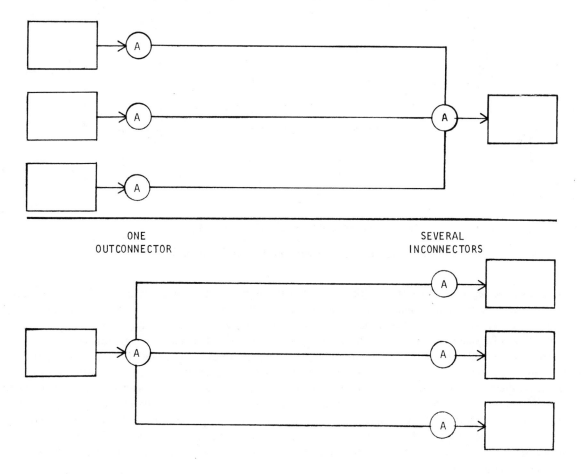

ONE
OUTCONNECTOR

SEVERAL
INCONNECTORS

FIGURE 2.4. Connector Symbol Examples

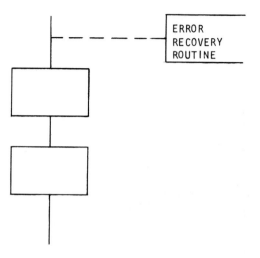

A HEADING OR BEGINNING OF A SUBROUTINE

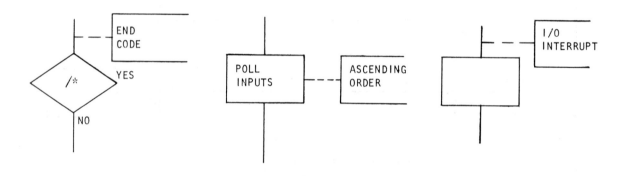

DETAILS OF A DECISION, INTERNAL PROCESS OR I/O OPERATION

FIGURE 2.5.   Annotations Symbol Examples

In many cases the begin/end, start/stop, initiate/terminate, or interruption/resumption are significant enough to warrant special attention and eliminate the question of locating the starting point and where the operation ceases.

In software and hardware operations, the terminal symbol expresses terminal conditions: beginning and ending. The word START inside the terminal symbol can imply a lengthy process such as start up, initialization, or read in. The start message symbol is always drawn with only one flowline (output), emanating from it in the direction of activity. By definition, there is one "start" or beginning position, and its location on the flowchart is worth noting.

Similarly, the word STOP enclosed by the terminal symbol can imply a lengthy process such as clean up details, terminating operations, or write output operations. The use of an input flowline to the symbol, but no output flowline, serves to indicate a period of inactivity or a quiescent state. In this manner, a halt, delay, or an interrupt can be indicated. These occurrences may be randomly distributed throughout the flowchart. The only means of identification is a unique symbol for this data processing operation. The stop oval symbol is used for this purpose, since all flowlines may not terminate at one common point.

In the present Standard (ANSI X3.5–1970), the meaning of the terminal symbol label has been extended to include the word interrupt, and the symbol is used to perform a supplementary function. In conjunction with the striped symbol notation, the terminal symbol performs the basic function of first (start) and last (stop) symbols of a detailed representation (see Figure 2.6). This representation is not in conflict with a predefined process symbol which identifies an operation(s) whose details are not represented in the same set of flowcharts.

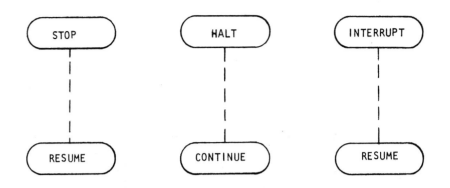

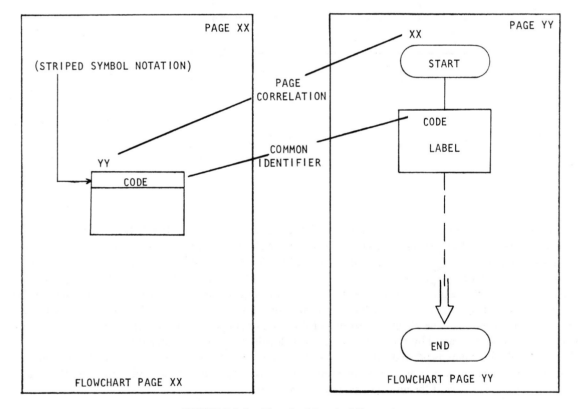

FIGURE 2.6. Terminal Symbol Examples

PARALLEL MODE SYMBOL

In the past, parallel processing was a rarity due to hardware technology and configuration. Currently, with hardware technology advances and software innovations, parallel processing is commonplace. Today concurrent or simultaneous software operations include, but are not limited to, multiple processing (fixed and variable tasks), microprogramming, virtual memory (CPU–I/O concurrency), associative memory (synchronous system), on-line timesharing and real time processing.

The parallel mode symbol outline consists of two parallel lines of any equal length, spaced sufficiently apart so as not to be misinterpreted with any other pair of lines within the same flowchart. The symbol is shown in the horizontal and vertical configuration.

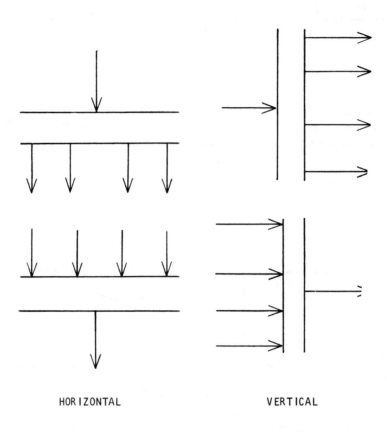

HORIZONTAL　　　　　　　　　VERTICAL

As shown above the parallel mode symbol, by itself, does not show any correlation between the input and output (entrance and exit respectively) flowlines. This problem always exists with this symbol usage. Unless adequate information is included, a mental analysis is required for flowline association. Some of the problems associated with this symbol usage are given here. From a processing point of view, the information for a given path may be on a time basis (multiplexing) or on a bit position basis. Further complexity can occur when information flow encompasses several paths and is on a time, block or record, or bit position basis. This difficulty is not reduced with additional annotations. The merging or sorting of flowlines (many in and one out or one in and many out, respectively) may be best served by the details given under flowlines in the basic category.

There are other problems associated with the use of the parallel mode symbol. Since concurrency is expressed here, numerous operations may begin at the same time and likewise terminate at the same time. Here the parallel mode symbol functions as start (begin) and stop (end) terminal operations.

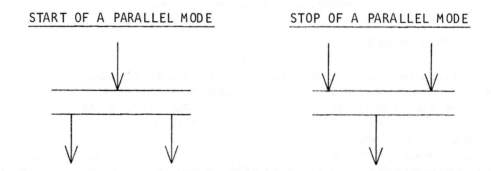

START OF A PARALLEL MODE　　　　STOP OF A PARALLEL MODE

A simple set of flowchart symbols has been developed, and it is appropriate to use the vocabulary at hand to do some problems. This primitive set permits all possible logical (sequential, repeat, and branch) operations to be flowcharted, but the drill exercise is only to develop familiarity with the flowchart symbols. Flowcharting skill will be developed progressively by this means. The remaining flowchart symbols encompass an area that includes hardware device representation and their data structures, software operations, and the integration of software and hardware operations.

EXERCISE

1. Using only the decision symbol, prove the selection of the largest of three numbers A, B, and C. Hint: Do the problem using only three decision symbols.

    ANSWER

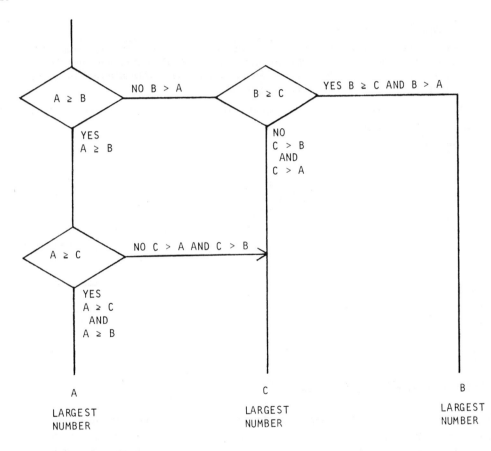

2. Do Exercise 1 for the smallest number of A, B, and C. What is the flowchart logic difference between the two exercises? (Hint: What can be said about the procedures?)

    ANSWER

    Both flowcharts are the same, Topologically the expression for $X > B$ and $X < B$ can be expressed as XRB where R is defined as the relationship operator or function.

3. In Exercises 1 and 2, the output could be combined as a single output. Why is it not possible for several numbers to be present at the same time?

    ANSWER

    The operation is sequential and only one path can be active at any one time.

4.   A parallel operation is shown here. Associate a code with each output arithmetic operation.

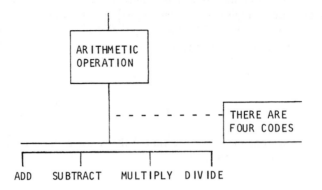

ANSWER

The single input and multiple output processing requires additional information to associate each code for each arithmetic operation. None is given. Therefore, the best interpretation of the above operation is that the arithmetic selection is accomplished simultaneously. If only one arithmetic operation is permitted at any one time, the process of selection (e.g., time) is the same regardless of the arithmetic operation.

5.   The output of a process is to be distributed to three other processing operations. Graphically present continuity of data flow.

ANSWER

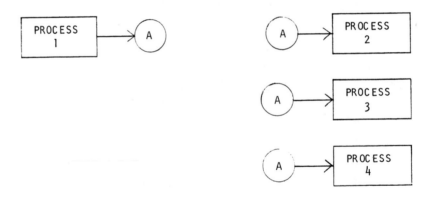

SYSTEM HARDWARE SYMBOLS

There is a certain amount of interfacing to be done between the computer and its external environment. Information must be prepared and converted into a form that is compatible with computer entry. Similarly a reverse procedure is necessary to convert the computer output into a form for human understanding (see Figure 2.7). In addition, there may be a continual dialogue between the input/output device and computer. Known as "handshake routines" such running conversations are sometimes necessary to accommodate the equipment operation and its limitation. In many cases, the utility programs are supplied along with the equipment to accomplish the I/O routines per equipment device. The flowchart symbols presented for this area cover the manual operations and functions for computer entry. There are symbols that denote the medium on which the information is stored or the manner of handling the information or both. In later chapters, some of these operations are further detailed.

If no specialized symbols exist, the input/output symbol is used.

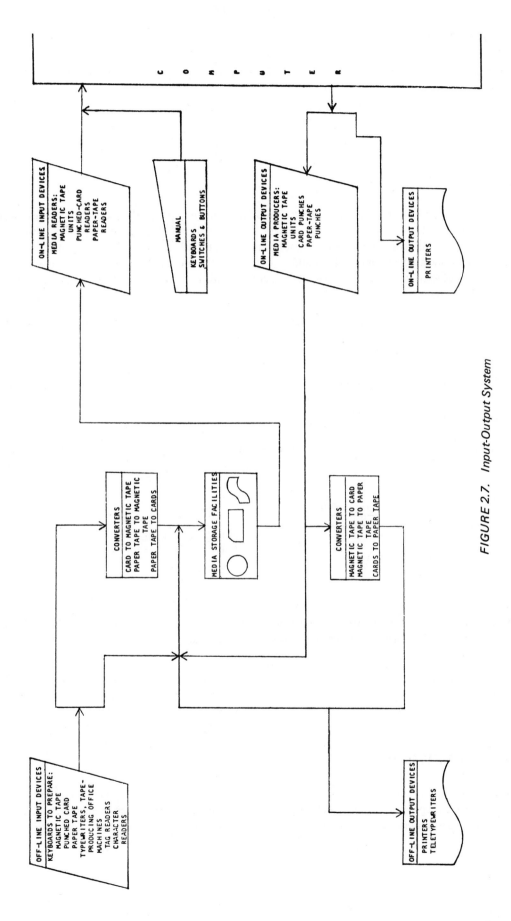

FIGURE 2.7. Input-Output System

INPUT/OUTPUT

The symbol shown below represents an input/output (I/O) function or operation of an I/O device.

    SYMBOL SHAPE   MEANING

                                AVAILABLE INPUT
                                DATA OR OUTPUT
                                PROCESSED
                                INFORMATION

Just as the terminal symbol is necessary to locate the starting and end point, the I/O symbol serves the purpose of stating that the input (reading) operation is necessary for information to be available, and the output (writing) operation is the end or task completion. This symbol, when used, is basic in a data processing unit (CPU), peripheral device, or any associated equipment. It should be noted that the symbol describes an input or an output function without necessarily denoting any particular method. Addition of a legend inside the symbol can add more specificity, as shown here.

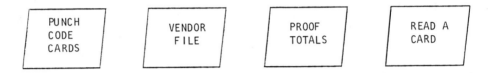

MANUAL OPERATION SYMBOL

The manual operation symbol in a flowchart informs the reader that automatic operations are not included at the point in question, and an operator is required. Hence a process requiring a human being is identified by the symbol shown:

    SYMBOL SHAPE   MEANING

                                MANUAL OFF-LINE
                                OPERATIONS.

                                CLERICAL
                                OPERATION.

The X3.5 Standard is very restrictive in defining this symbol. The process performed should be off-line and geared to the speed of the human beings, without using mechanical aid.

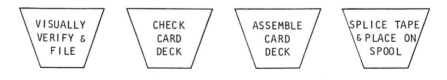

MANUAL INPUT SYMBOL

The manual input symbol is more definitive than its two predecessors, Input/Output and Manual Operation. This symbol, shown below, represents an I/O function in which the information is entered manually at the time of processing by means of on line keyboards, switches, pushbuttons, readers, plugboards, etc.

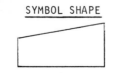

    SYMBOL SHAPE   MEANING

                                INPUT SUPPLIED
                                TO THE COMPUTER
                                ON-LINE.

This includes interruption or insertion of data or equipment by manual manipulation of any device in current use. On the basis of the expanded definition, the manual interruption is applicable to Input/Output operations and not just input operations exclusively. Generally, the manual operations are always inputs, but their results can alter and modify the computer operations in terms of output.

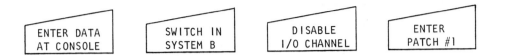

At this point, it is appropriate to distinguish and clarify the previous three symbols:

1. Input/Output Symbol
2. Manual Operation Symbol
3. Manual Input Symbol

The first one is self-explanatory. It primarily states the input and output requirements. The second item is restricted to manual operations off-line without mechanical aids. The third item details the means, technique or device to be used to obtain manual operation in the input/output area.

DISPLAY SYMBOL

Display is a visual representation. No restrictions are defined by the ANSI Standard X3.5. Displays can include indicators (generally lamps, switches, and dial positions), alphanumeric readouts, meters, printers, plotters, and CRT graphics. Obviously, the symbol is used exclusively in the output area, just as the manual input symbol is restricted to the input area. The symbol shown below represents a function in which information is displayed for human use.

SYMBOL SHAPE   MEANING

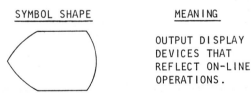

OUTPUT DISPLAY
DEVICES THAT
REFLECT ON-LINE
OPERATIONS.

This symbol represents information displayed for operator interpretation at the time of processing by means of on-line visual aids. Like the manual input symbol the display represents an on-line operation. The two symbols, display and manual input, close the loop from output to input to complete the feedback path in a control system configuration (see Problem 4).

STORAGE SYMBOLS

Before continuing on to the next two symbols, we will define the terms "on-line" and "off-line". "On-line" means under direct control of the central processing unit. A process or task is performed "on-line" or an item of computer peripheral equipment operates "on-line" if under direct control of the CPU. In a complex hierarchy configuration, a subsystem is designated as the master or executive control of the total system and has its peripherals under direct control to be on-line. "Off-line" means not under control of the central processing unit per se. A device such as a tape station (digital magnetic recording facility) can be operated off-line even though it is in the immediate vicinity of the CPU. Conversely another tape station, located at a distant satellite or remote installation, can operate timeshared with the CPU for on-line operation.

ON-LINE STORAGE SYMBOL

The symbol shown below represents an I/O function associated with external memory of the CPU for on-line operation purposes.

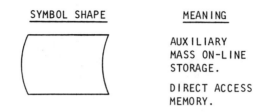

SYMBOL SHAPE

MEANING

AUXILIARY
MASS ON-LINE
STORAGE.

DIRECT ACCESS
MEMORY.

Devices for mass storage of information, both on-line and external, include magnetic drums, discs, tape and strips, automatic magnetic card and punched card systems. By definition automatic microfilm chip and strip systems are included. Numerous slow speed peripheral devices are included, but are assigned to off-line operations because of time consumption.

OFF-LINE STORAGE SYMBOL

The symbol shown below represents any off-line storage of information, regardless of the medium on which the information is recorded.

SYMBOL SHAPE

MEANING

AUXILIARY
MASS OFF-LINE
STORAGE.

Legends, as shown here, allow additional description of the off-line storage purpose, method, or medium (hardware).

INPUT/OUTPUT (HARDWARE) SYMBOLS

The following symbols are specific and designate particular equipment. These I/O symbols may represent the I/O function and, in addition they can denote the medium on which the information is recorded, the manner of handling the information, or both. Each symbol represents a complete subsystem or subroutine operation by virtue of the symbol profile. The picture word alone conveys the information so succinctly that written descriptions are usually not necessary.

PUNCHED CARD SYMBOL

The symbol shown below represents an I/O function in which the medium is punched cards, including mark sensing cards (character recognition equipment), partial cards, stub cards (marked scan), or any type of hard card copy and flexible cards as a medium of information storage.

When this symbol is used, it defines an established processing operation. Presently there are formalized procedures related to each of the various categories designated by this symbol. The physical size, coding, or format dictate the system parameters and configuration. The versatility and flexibility are obtained by uniformity and compatibility of software and hardware among users of this medium. As noted here, the punched card symbol is an I/O operation and has a hardware connotation. The punched card signifies more than a card as an input storage medium. The total information package to the programmer includes the coding (ASCII or other codes), the data format and structure per card, and for chain (data continued on the next card) card inputs. All this information can be conveyed with the use of a single flowchart symbol, the punched card. An advantage of the punched card is that it can contain both digital information for computer entry and text information for the human user. A logical record and the physical card (e.g., 80 or 96 columns) can have the same data capacity. Therefore, data on a per card basis can be modified, deleted, or inserted in a card deck without impact on adjacent cards.

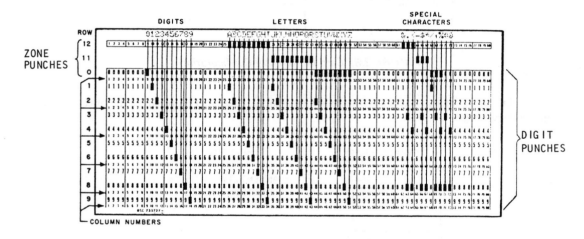

Each numeric, alphabetic, and special character can be represented in Hollerith punched hole code by one, two, or three holes punched in a column as shown.

SPECIALIZED PUNCHED CARD SYMBOLS

Since punched cards provide considerable flexibility and versatility in generating, modifying, updating, correcting, and assembling a program, the X3.5 Standard has three symbols related to punched cards alone. In addition to the single card symbol already shown, specialized representations are available showing a punched card as part of a deck of cards or as part of a file of cards. Previously, the single punched card symbol served for all these functions. The multiple card symbol offers possible card-by-card labelling and use of legends to describe more fully the function or process being depicted.

SYMBOL SHAPE   MEANING

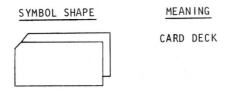

CARD DECK

As shown above, no meaning is attached to the deck of cards. It is just a collection of punched cards to be processed (input) or just a collection of cards thay may result from a compiling or merging operation (output) or even a representation of a copying operation. Appropriate legends can augment the symbols however, and a breakdown of each card can be easily represented and detailed as illustrated in Figure 2.8. The card representation can be enlarged as necessary to contain the text material.

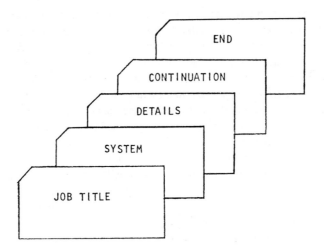

FIGURE 2.8. Card Multiplicity Configuration

Some overall meaning can be attached to the following symbol, because it represents a file of cards (i.e., a collection of related cards).

SYMBOL SHAPE   MEANING

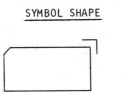

AN ENTITY
RELATED PUNCHED
CARD

A label can be addressed or attached to this collection of cards. The symbol construction is identical to the perivous symbol with the exception of a deletion of line segments to obtain the above outline. An assembly of related punched cards creates a card stack (deck of cards) and is represented by the deck of cards symbol. A hybrid notation of multiplicity and deck of cards is shown in Figure 2.9.

PUNCHED TAPE SYMBOL

Next is the symbol for an I/O device or function using punched tape (paper or film) as the data storage medium. Punched tape serves much the same purpose as punched cards and can be used as either an input (read) medium or an output (punch) medium.

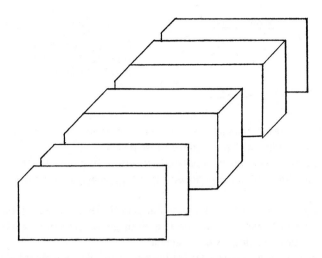

FIGURE 2.9.   *Card Multiplicity Hybrid Configuration*

SYMBOL SHAPE	MEANING
	SPOOLED PUNCHED MEDIUM FOR I/O OPERATIONS

This symbol is associated with teletype I/O, paper tape punching equipment, typewriters with punching equipment attachments, and manual preparation of the punched tape medium. There are standards of tape dimensions, and coding conventions for five, six, seven, and eight levels (channels) on the tape. This medium is extensively used in numerically controlled machines. Any tapelike perforated strip may be adequately represented by the punched tape symbol. Strips containing information in bit form (transparent and opaque) may be included even though the coding is optical in nature. The punched tape symbol denotes the total input/output service functions required for this storage medium in digital data processing configuration.

The punched paper tape is a continuous medium, and lacks some of the flexibility of a punched card deck. Data format must be defined so that the beginning and end of each parameter (or word) can be blocked (or formed). The sprocket channel insures each code line is read correctly. As a slow speed device, punching or reading, the paper tape device can be moved and stopped on a single code line. Program modification is not as simple as the punched card. Splicing is used more often for paper tape repair than for insertion of new data.

The segment of punched tape below shows a typical punched hole code for numeric, alphabetic, and special characters.

MAGNETIC TAPE SYMBOL

Magnetic-tape recording of digital data is so standardized in format and coding that interchangeability among different tape recording equipment and computer complexes is common practice. The symbol shown below represents an I/O function in which the medium (magnetic tape, strip or loop) is processed by a magnetic tape station (MTS).

SYMBOL SHAPE   MEANING

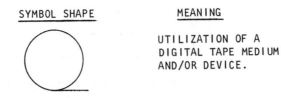

UTILIZATION OF A
DIGITAL TAPE MEDIUM
AND/OR DEVICE.

This symbol is not used to represent magnetic recording media other than tape (e.g., cards, drums, or discs) simply because the MTS is not a prerequisite device for those media.

The magnetic tape symbol denotes the total input/output service functions for this storage medium in the digital data processing configuration. Like the punched paper tape, magnetic tape is a continuous and flexible medium.

If desired, one could manually read and check the digital data (holes punched) on punched cards and punched paper tape. This is not possible with magnetic tape or data on magnetic discs or magnetic drums. Selective erasure and writing over the same physical locations is not practical.

In tape-oriented processing, updating consists of reading an entire file (one record at a time) and writing an entirely new one. In disc-oriented processing, updating consists of re-writing only those records that require change. No new file need be created, if the re-writing operation permits such an updating arrangement (file structure).

Tape drives process data in a sequential access method. (Sequential processing simply means that records are read or written one after the other.) Therefore, records stored on magnetic tape are normally arranged sequentially on the basis of some type of "keying" arrangement (e.g., item number, time word, date, employee number, or alphabetic listing).

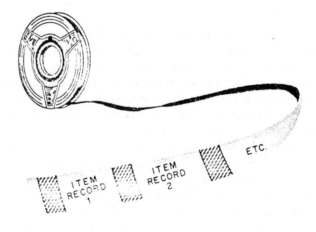

Numeric and alphabetic characters are represented by magnetized spot patterns as shown in the simplified drawing of a tape segment below.

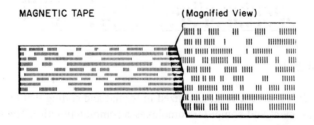

Alphanumeric information is recorded on tape just as songs are recorded on tape, one after the other. If the data from 2000 sales slips is recorded on a tape, the magnetized spot patterns representing sales slip number 1000 can be read and processed only after all sales slips from number 1 through number 999 are searched sequentially.

MAGNETIC DRUM SYMBOL

The flowchart symbol representing magnetic drum recording or playback is a recent addition to the Standard (ANSI X3.5–1970). The symbol depicts the cylindrical shape of the drum used in the actual device. Ignoring the applications, the location of this device is typically that of an internal (or integral) part of a computer, and we are representing it as an I/O function. In the past and even today, we have drum equipped computers in which the device is definitely not peripheral to the computer. As a semirandom addressing device, the drum is generally associated as an on-line internal memory function. Based on the field of a computer word, where coding is used to identify internal or external addressing, the drum is identified as an on-line device. In file retrieval systems, the drum may comprise the total system storage. In large systems, the drum may perform an intermediate storage function, such as buffering between a slow-speed peripheral device and the high-speed internal memory and vice versa. When mass storage is specified, the following outline is available.

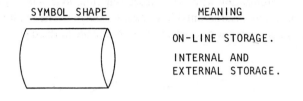

```
SYMBOL SHAPE                    MEANING

                               ON-LINE STORAGE.

                               INTERNAL AND
                               EXTERNAL STORAGE.
```

The drum stores data on its surface magnetically in the same manner used for magnetic discs. The data is serial by track, and several tracks can be read or written simultaneously. A typical magnetic drum layout is shown next. The data layout has fast access bands and main storage bands. The fast access bands permit data accessibility more frequently than once per revolution.

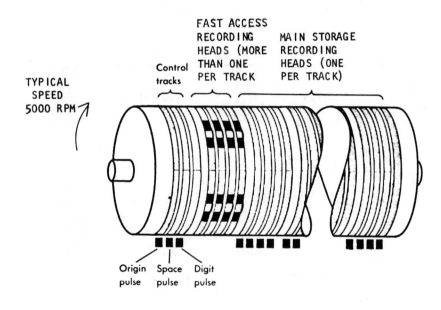

MAGNETIC DISC SYMBOL

Although disc storage entered the processing field after magnetic tapes and drums, its rate of growth indicates that disc storage will probably overshadow its predecessors. Again it is difficult to classify the disc device as an I/O function. Normally, discs are directly addressable and are an integral part of the computer installation. The programmer, seldom if ever, has the disc under his control, except via a disc controller. Again, if I/O channels are addressable and equipment shared, then we may classify the disc as an I/O function. The disc, as a storage function, is classified along with magnetic tapes and magnetic drums as mass storage devices. Functionally, the disc has the advantage of the shortest access time while retaining the same transfer rates. Where disc storage or disc processing is implied, the following symbol is applicable.

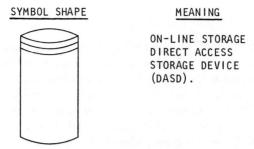

SYMBOL SHAPE   MEANING

ON-LINE STORAGE
DIRECT ACCESS
STORAGE DEVICE
(DASD).

In disc processing, the record formats are quite similar to tapes, and fixed and variable record length data formats are shown in Figure 2.10. The rigid medium and control signals permit disc operation to be scheduled in advance and to be executed in synchronization with other processing functions. Although speed variation (spindle rotation) does exist, it does not pose any insurmountable software problems. The random access, storage volume, and rate of throughput provide system operation versatility not obtainable in the digital magnetic tape and magnetic drum operation.

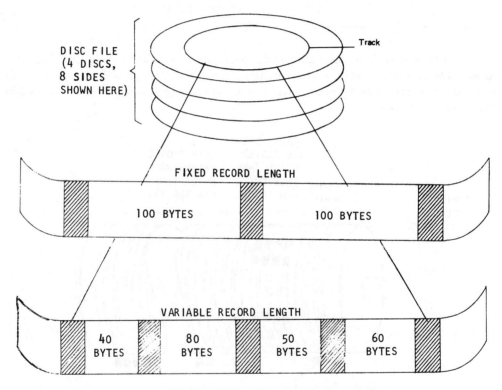

FIGURE 2.10.   Disc Data Format

CORE SYMBOL

The core symbol outline looks like a sketch pad that is used primarily for temporary results, and that is exactly its application. Generally, magnetic cores (Coincident Core Magnetic Memory – CCMM) are associated with the main or internal memory of a CPU. This does not exclude the use of core memories external to the CPU. Core memories are being used as add on memories and applications of buffering and formatting external to the internal memories. In any event, the core symbol is a very specific type of hardware and application. The last three flow-chart symbols (tape, drum, and disc) and the core symbol are all representations of the configurations of actual magnetic devices and subject to obsolescence in the event of the appearance of competitive technologies. Rather than be subjected to everchanging technologies, these symbols may be elevated to higher levels of abstraction and made to conform to generic descriptions such as file, mass storage, and intermediate or temporary storage. At present, the core memory (and possibly solid state cells) is the fastest access device among all four. The core symbol outline is shown here.

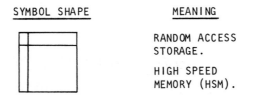

SYMBOL SHAPE   MEANING

RANDOM ACCESS
STORAGE.

HIGH SPEED
MEMORY (HSM).

DOCUMENT SYMBOL

The symbol shown next represents an I/O function in which the medium is a document. Some examples of typical documentation, such as might be represented by the document symbol, are shown on the next page.

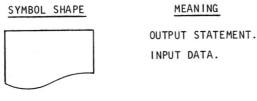

SYMBOL SHAPE   MEANING

OUTPUT STATEMENT.

INPUT DATA.

Whereas the punched card, magnetic tape, and punched tape symbols represent input and output peripheral devices, the document symbol may need additional clarification when it is used. It is well established that the output function, represented by the document symbol, includes such devices as page or ledger printers, typewriters, and possibly column printers. Film copy of reports, ledgers, and documents might be represented by this symbol solely on the basis of its paper copy denotation. However, as an input source, the document symbol still represents paper copy and reports of all varieties including source documents and ledgers. The preparation, translation, and conversion of paper copy into a fully integrated automatic data processing system (ADP) is not so simply stated by one symbol. The complexity of document entry into an ADP requires a number of unique paper handling procedures and specified equipments that may include any of the previous symbols for this category.

The flowchart symbols covered in this section are primarily used to define and configure a system or solution without going into detail. Figure 2.11 presents an overview of a complex data processing configuration using the symbols just presented. For simplicity, the core symbol, disc symbol, and drum symbol are not shown specifically, but are included in the rectangle labelled "computer processing." The figure sets forth not only the hardware "hook up," but a general introduction to the system definition and description of capability and complexity.

Figure 2.12 presents a comparison of the performance of magnetic devices in terms of capacity, operational speed, and data accessibility. The mass storage units are being assigned key roles in timesharing, communication, and on-line processing operation. The merit and justification for the wide range of devices should be apparent to the reader. Going from output to input, there is a range of data accessibility from low speed to high speed and from high speed to low speed for reverse direction. The equipment performance or any limitation must be fully understood prior to programming these devices. In general, the I/O routines are quite difficult to generate, test, and debug.

DOCUMENT SYMBOL EXAMPLE

INPUT DATA CODING

```
STB   SSAV      FOC BO
JST   SIN8
IAB
MPY   AMPC
STA   ELOF      ELEVATION OFFSET FOC BO
LDA   SSAV      COMPUTE AZIMUTH OFFSET
JST   COS8
IAB
MPY   AMPC
STA   AZOF      AZIMUTH OFFSET FOC BO
LDX   SOUR
JMP   *+1, 1
JMP   SDUN      CIRCLE SCAN PT. Done
JMP   SLSD      Side Lobe Circle.Scan-PT. Done
```

DOCUMENT SYMBOL EXAMPLE

OUTPUT STATEMENT - PROGRAM LISTING

							COMPUTE OFFSETS FOR CIRCLE SCAN SUPER-POSITION
14155	0	74	14177	CIRC	JMP	ICIR	INITIALIZE CIRCLE SCAN
14156	0	67	14752		IRX	CSCN	COMPUTE POINT ON CIRCLE SCAN B23
14157	0	57	00000		IAB		
14160	0	34	14753		MPY	PT20	B23*B0=B23 1/PTS PER REV.
				*			THE B-REGISTER NOW CONTAINS THE POSITION OF THE
				*			RADIUS VECTOR IN FOC BO
14161	0	03	14754		STB	SSAV	FOCBO
14162	0	27	14532		JST	SIN8	
14163	0	57	00000		IAB		
14164	0	34	14755		MPY	AMPC	
14165	0	05	14733		STA	ELOF	ELEVATION OFFSET FOC BO
14166	0	24	14754		LDA	SSAV	COMPUTE AZIMUTH OFFSET
14167	0	27	14534		JST	COS6	
14170	0	57	00000		IAB		
14171	0	34	14755		MPY	AMPC	
14172	0	05	14732		STA	AZOF	AZIMUTH OFFSET FOC BO
14173	0	56	14756		LDX	SOUR	
14174	1	74	14175		JMP	*+1,1	
14175	0	74	12542		JMP	SDUN	CIRCLE SCAN PT. DONE
14176	0	74	13224		JMP	SLSD	SIDE LOBE CIRCLE SCAN PT. DONE

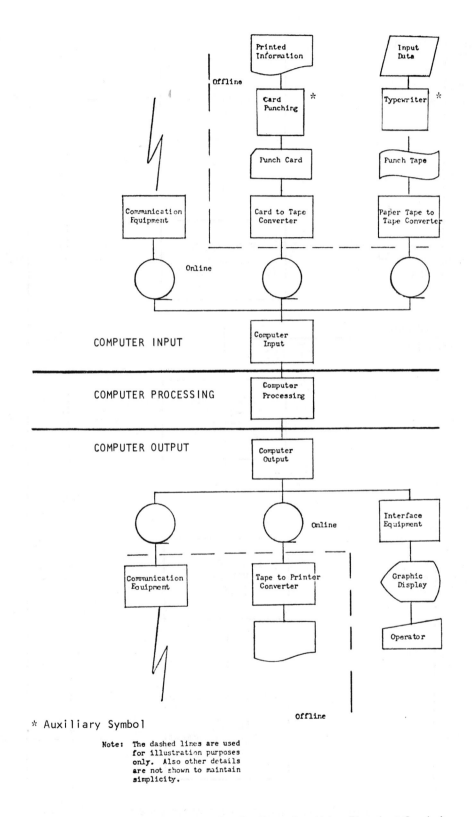

COMPUTER INPUT

COMPUTER PROCESSING

COMPUTER OUTPUT

* Auxiliary Symbol

Note: The dashed lines are used for illustration purposes only. Also other details are not shown to maintain simplicity.

FIGURE 2.11 Complex Data-Processing Configuration Using Flowchart Symbols

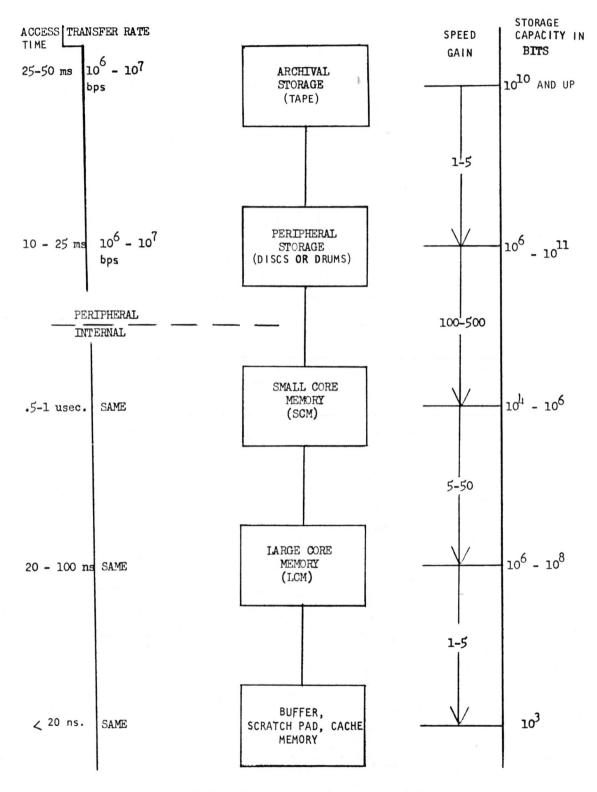

FIGURE 2.12 Comparison of the Performance of Magnetic Devices

EXERCISE

6. What symbol can be used to supply more descriptive information than the standard input/output symbol concerning input/output operations?

ANSWER

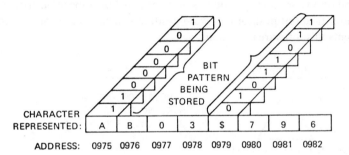

7. Both data and instructions are stored in memory in binary code. An alphanumeric character is stored in eight cores and is represented by eight bits. When stored in memory, each character has an addressable location as shown in figure below.

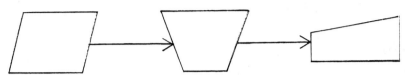

What is the address of B? What does the number 0982 represent? If the symbol $ is used in a program statement, where would it be found?

ANSWER

0976; address of a memory location; 0979

SYSTEM SOFTWARE SYMBOLS

System software symbols are evolving in a manner similar to the evolution of the system hardware symbols just described. These symbols can be seen in many recent software system proposals and presentations. As yet, the Standard has not been finalized in this area. There are a few symbols available in an elementary form for such applications and probably more will follow at a later date.

SPECIALIZED (FILE) PROCESSING OPERATIONS

Four basic software processing operations are examined here. These symbols are applicable for computer and noncomputer operations. For illustrative purposes, the software operations is discussed here in terms of typical file processing, with magnetic tape serving as physical and logic records. The four flowchart symbols are classified as system symbols, since they can cover a broad area without specific details on record formats and therefore do not disclose the programming complexity. However, they identify and highlight system requirements. Generally speaking, the four processing operations cover a wide latitude, and file processing is just one method of implementation.

MERGE SYMBOL

A typical merge operation results in the combining of a master file and an input file to produce an updated master file. Thus, a single ordered file is formed by combining two or more ordered files. The form of the items is not changed during a merging operation. The file may comprise items associated with inventory control, invoice control, and payroll, to mention a few. The symbol shown represents this software operation.

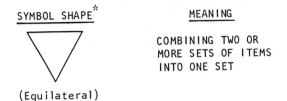

SYMBOL SHAPE*   MEANING

COMBINING TWO OR
MORE SETS OF ITEMS
INTO ONE SET

(Equilateral)

EXTRACT SYMBOL

The merge process is the combining of multiple sets of input items to achieve one ordered output set, and the extracting process is the exact reverse. Because of this the extract symbol is drawn as a merge symbol rotated 180 degrees about the horizontal axis. Also, the hardware implementation has a single input (set of items) and the output represents the removal of one or more specific sets of items based on a specified criterion. The criterion may be a key word, a computer field (bit position), or word position in a data block, or any esoteric term. The symbol shown below represents this software operation.

SYMBOL SHAPE   MEANING

REMOVAL OF ONE OR
MORE SPECIFIC SETS
OF ITEMS FROM A SET.

(Equilateral Triangle)

SORT SYMBOL

When a sort operation is performed, a set of items is arranged in a particular sequence. Pictorially speaking, from its outline construction, the sequence of extraction and merge processes are performed. In actual practice, the sort comprises this two-step sequence. Initially, a group of items is generated (extraction). Then the groups are examined using a set of characteristics to combine and obtain a resultant overall ordered set. The symbol shown below represents this software operation.

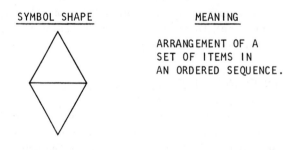

SYMBOL SHAPE   MEANING

ARRANGEMENT OF A
SET OF ITEMS IN
AN ORDERED SEQUENCE.

COLLATE SYMBOL

The collate operation can be described as merging and extracting in sequence. In a previous example of a master tape and input tape, the collate operation is analogous to an extraction process after the files have been updated (merged). In essence, a collate operation consists of the comparing and merging of two or more similarly ordered sets into another ordered set(s). The symbol shown below represents the software operation of merging with extracting.

*The above symbol and off-line storage symbol both use the same cutout in most drawing templates, but the off-line symbol has an additional line segment.

SYMBOL SHAPE

MEANING

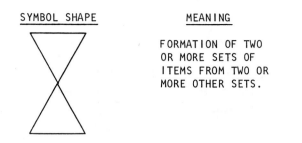

FORMATION OF TWO
OR MORE SETS OF
ITEMS FROM TWO OR
MORE OTHER SETS.

It is appropriate at this time to illustrate each of the operations (merge, extract, sort and collate) using a deck of cards.

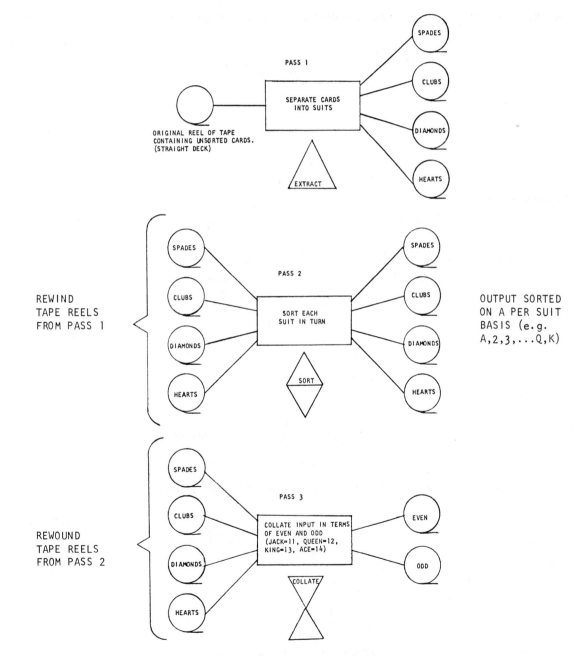

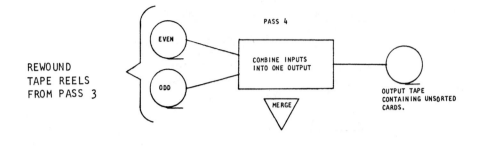

REWOUND
TAPE REELS
FROM PASS 3

PASS 4

EVEN

ODD

COMBINE INPUTS
INTO ONE OUTPUT

MERGE

OUTPUT TAPE
CONTAINING UNSORTED
CARDS.

EXERCISE

8. Implement a merge operation using only magnetic tape and process symbols.

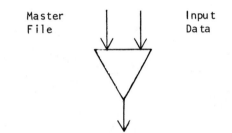

Master
File

Input
Data

Update Master File

ANSWER

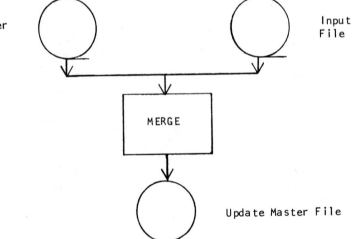

Master
File

Input
File

MERGE

Update Master File

9. Implement an extract operation using only magnetic tape and process symbols.

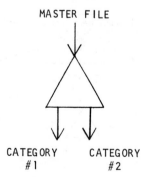

MASTER FILE

CATEGORY
#1

CATEGORY
#2

ANSWER

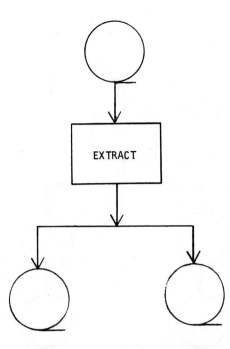

10. Implement a sort routine using only magnetic tape and process symbols.

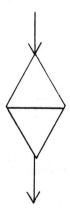

HINT

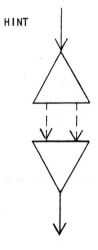

ANSWER

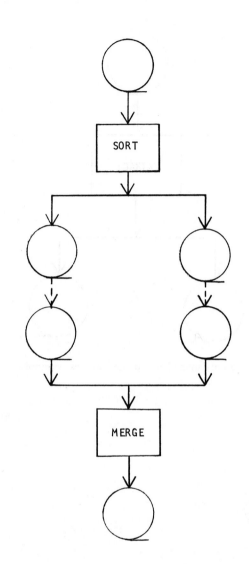

11. Implement a collate operation using only magnetic tape and process symbols.

ANSWER

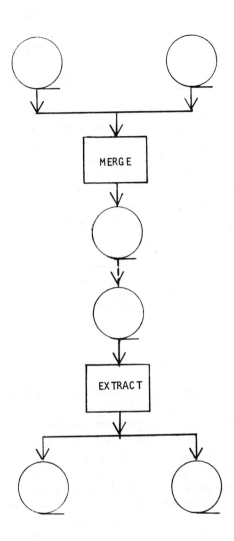

12. In problems 8 and 10, is the input operation sequential or parallel? Draw a flowchart of problem 8, and explain.

    ANSWER

    Sequential operation. The master file record is read first. The update file is read next. A decision is made to output one of two records. If the update file record is outputted, then the next update file record is read. If the master file record is outputted, then the next master file record is read. The process is repeated until both files are exhausted.

SPECIALIZED PROGRAMMING OPERATIONS

In documenting software system configurations and programming operations, a few flowchart symbols are available and others are to be created and incorporated into the set of standards. Two flowchart symbols are presented here and several potential areas are examined in the next sections on specialized operations.

PREDEFINED PROCESS SYMBOL

Another broad and general purpose symbol is the predefined process symbol. The symbol shown here represents a named process consisting of one or more operations or program steps that can be specified.

SYMBOL SHAPE   MEANING

GROUP OF
OPERATIONS-
ITERATION

SUBROUTINE

This symbol implies the existence of the subject manner as an entity. It may be currently available or may be constructed or generated. In any event, at the top level, the predefined process is acknowledged and the resulting details are available or assumed to be available. Any named process consisting of one or more operations or program steps so detailed can be inserted to replace or be substituted for this symbol. Just as the connector symbol can be used to connect different levels of flowcharts, the predefined process symbol can be used to relate two levels of hierarchy in a specific fashion. Succinctly, the predefined process symbol represents a group of operations detailed elsewhere or externally referenced. A typical example would have the enclosed text material calling for an external routine to the current program (listing).

PREPARATION SYMBOL

The hexagon symbol has been used in the past for a predefined process and to detail the information content at another flowchart level. However, the Standard (ANSI X3.5–1970) has chosen the vertical strip on each side of the process symbol for this function. Since the hexagon symbol is available for other considerations, it is being redefined and used as a preparation symbol. The name "preparation symbol" is misleading. The symbol is a pseudonym for program modification. When one considers preparation in the context associated with software and programming, the definition covers a wide range of applications. Preparation for on-line purposes may include all the preprocessing, preconditioning, and examination of raw data prior to its evaluation, verification, and application for computer processing. For off-line operation, the symbol may be used to represent all the manual paper processing of data prior to its reduction into a medium for computer entry. Invariably, there are a number of time consuming and costly preparation procedures to any computer entry. Yet the identification, classification, and tabulation of these items are of paramount importance to any ADP operation. The word "preparation" draws attention to these procedures and puts them in their proper perspective with other software functions and possibly hardware items.

The Standard defines the preparation symbol in terms of (program) modification of an instruction or group of instructions which change the program itself. In the past (but no longer) the symbol shown below was used to alert the user for this procedure.

PROGRAM MODIFICATION SYMBOL

In some cases the type or degree of program modification represented by the program modification symbol was not obvious to the reader using a flowchart and some accompanying text material was necessary. The above symbol, when not used as an operation, process, or function, was used as a means of annotation with a dashed line for the same purpose (program modification) as shown.

- - - - PROGRAM MODIFICATION ANNOTATION

In both cases, operation and annotation, this symbol usage is not consistent with established symbol standards. Instead, the program modification symbol is used to indicate that an instruction or group of instructions change the program per se (e.g., address modification, operation code modification, modification of index registers, setting or clearing, and initialization). This type of program modification is an integral part of a complete program.

The symbol shown below represents this software operation.

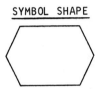

SYMBOL SHAPE

MEANING

PROGRAM MODIFICATION.

PROGRAM PREPARATION.

This symbol permits identification of the following areas:

1. Program modification of a computer program
2. Mode selection or program modification
3. Process selection
4. Initialization
5. Base register selection or change*

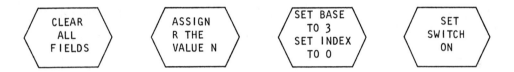

OTHER SPECIALIZED OPERATIONS

In planning, defining, and configuring a software system, it is necessary to formulate and present concepts and to document them. The formulation, evaluation, and tradeoff considerations that arise must be recorded and documented. The needs for communicating at this level are great, and any assistance of new flowchart symbols is appreciated. The current Standard is lacking in this area, but it is only a matter of time until a revision will be issued to cover this area. No symbols are presented here, but a few software operations and parameters are suggested and a few flowcharts are reproduced later.

Some of the specialized operations and parameters to be defined may include program modules (processing), tables, and other sundry details such as: common service routines; performance monitoring (PM) and fault isolation (FI); private and common data base areas. A top level or summary flowchart of a software system identifies the interface among the program segments and the data flow. Also innovations such as virtual memory and associated memory have added another dimension to software design and programming. In addition, microprogramming is another type of parallel or simultaneous software processing. Not only must these new operations be investigated to determine if present flowchart symbols are adequate, but if new symbols are necessary, what outline, form, definition, and conventions are necessary to completely implement them. Broad concepts depicted for system concepts determine the merits of software implementation and the first cut of system tradeoffs.

*A base register holds a specific reference address to which relative addresses (in machine instructions) are added to determine absolute addresses of storage locations.

SUPPLEMENTARY SYMBOLS

Supplementary symbols do not significantly contribute in any particular way to extend operations, functions, or processes that have already been enumerated. In fact they are auxiliary symbols available for limited application purposes.

AUXILIARY OPERATION SYMBOL

In a multiple computer/multiple processing configuration there are primary and auxiliary equipment and operations having procedural priorities based on rank, time, and problem solutions. The square symbol shown below represents an off-line operation performed by equipment not under direct control of the central processing unit.

SYMBOL SHAPE   MEANING

OFF-LINE SUPPLE-
MENTARY TO THE
MAIN PROCESSING
FUNCTIONS

When an operation is performed that complements or supplements a current or on line operation, or that can be executed by the primary equipment (generally at reduced system performance), the symbol above is used. Just as the square is a subset of a rectangle (or the rectangle includes the square with its outline), the auxiliary operation as represented by the square symbol can be considered a subset of a process as represented by the rectangular symbol. The process symbol is used if no specialized symbol is available and can always be substituted for any specialized operation.

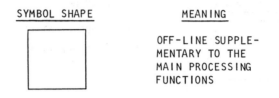

TYPE-WRITER   KEYBOARD TO TAPE   KEYBOARD TO DISC

COMMUNICATION LINK SYMBOL

The communication link symbol represents the transmission of information from one location to another via communication lines. The communications links may be by cable and/or radio methods. The communication link symbol implies the signal itself (not the information content) undergoes a conversion process for transmission compatibility purposes only. This technology invokes modulation (or encoding) and demodulation (or decoding) without any data processing involved. If any additional processing of the information content is used, it should be "broken out" and defined by the remaining flowchart symbols. The symbol shown below represents an I/O function in which information is transmitted automatically from one communication terminal to another.

HORIZONTAL   VERTICAL

Unless otherwise indicated, the direction of flow is left to right and top to bottom. Open arrowheads are necessary on symbols for which the flow opposes the above conventions. An open arrowhead may be used on any line whenever increased clarity will result.

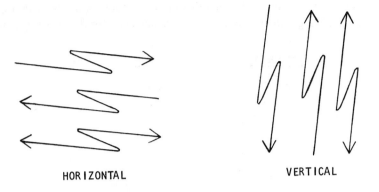

HORIZONTAL   VERTICAL

NONSTANDARD SYMBOL PRESENTATION

The standard symbols already described are adequate for most flowcharts, and they represent a consensus of what is being used. Additional unique symbols for input, output, and substitutions could be added, but special symbols are not essential. The frequent use of an item can be supplanted by assigning a special symbol rather than constant use of annotations or text material. As the technology changes, some will be removed while others will be added. Consequently, the basic symbols remain while the special symbols are current with the changing technology. The object being represented will be depicted in many ways, and numerous symbols will result. There are a number of devices and systems that can be assigned to this category. They may be classified on the basis of character recognition (bars, patterns, symbols, etc.) or the style of page or document (preprinted or printed). The medium form or transducer operation may be a method of classification. In any event, the proliferation of devices and systems will continue, and some will be standard comparable to the punched card, punched tape and magnetic tape systems. At this time, a flowchart symbol, its definition and conventional use for the above peripheral operations must await promulgation by the Standards Committee. A few are reviewed here. It should be noted that there are two areas of input information techniques, namely: optics (light) and voice. One is in practice, and one is still in research and development respectively.

OPTICAL INPUT (OCR, MICR, CIM) *

There are a number of optical readers and character recognition equipment available that serve as data inputs. The optical input symbol is the combination of the diamond shape (decision symbol) and circle as shown below:

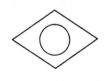

OPTICAL INPUT SYMBOL

OPTICAL OUTPUT (COM) **

The optical input symbol could be used to designate the optical output operation just as in the case of the punched card, punched tape, magnetic tape, and document flowchart symbols. However, there is a wide assortment of

*OCR, Optical Character Recognition
 MICR, Magnetic Ink Character Recognition
 CIM, Computer-Input-Microfilm
**COM, Computer-Output-Microfilm

equipment and operations in which a common flowchart symbol would not be appropriate for the various ways of recording the information. The most common optical output method is to display the data on a cathode ray tube and then photograph it. Another approach is to have an electron (or laser) beam record directly on film. A third approach involves a matrix of fiber optics; light emitting diodes attached to the far ends of the fibers form a display on the face of the fiber optics assembly. Still another optical output form is holography. Because of all of these alternate methods, the following symbol may prove to be inadequate for flowcharting purposes.

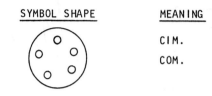

SYMBOL SHAPE   MEANING

                       CIM.

                       COM.

CREDIT CARD

The credit card denotes a media form and its application. The present flowchart symbol, a punched card, has become so specialized in its meaning (in terms of both hardware and software operations) that it has become inappropriate as a representation of the typical credit card. Credit cards are wallet size, usually, and they differ considerably from the usual punched card. They may carry legends that are printed, magnetically written, or even embossed. Computer entry systems for credit cards are yet to be finalized, but a card entry flowchart symbol would be a start in the right direction. The current punched card symbol has been used in a vertical orientation (as shown here) to denote a credit card data processing system.

CREDIT CARD SYMBOL

The notched corner, representative of a punched card, denotes a mechanical sensing and card alignment operation for computer entry. The credit card field is expanding and the cards are now used for telephone, transportation, reservation, and security systems, to mention a few applications. Here, both punched (optical) and magnetic cards are media forms.

VOICE COMPUTER ENTRY SYSTEMS

Intense research is being expanded to develop voice operated transducer devices, having digital outputs that interface with the computer. If the research proves successful, it will have a profound effect on the computer programming language category. This technique may be feasible, but it may prove to be too costly to be practicable and profitable for general use.

Graphic displays are expanding and encompassing fields such as automated education, computer aided design, and other visual presentaitons. The particular device used is the cathode ray tube (CRT). Special symbols denoting CRT use are shown below:

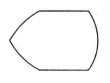

CONTROL CONSOLE   CRT DISPLAY
      OR
 CRT DISPLAY

Up until recently, the CRT device and its integration into the computer field were adequately covered by the current set of flowchart symbols. However, the proliferation of this device as a terminal device both for input and output operations now warrants re-examination and evaluation. The variety of applications that surround the CRT include it as a TV and "scope" monitor, and electronic measurement device (time, waveform, voltage, and current) may require a more proficient way of defining a hardware and software operation.

OFF PAGE CONNECTOR

The off page symbol is a special purpose symbol used to designate exit from or entry to another page. This symbol distinguishes between flowline continuity within a page and among pages. By definition, connector identification reference does not include symbol numbering and page identification and possibly location within a page to reduce searching time. Identification and referencing is covered in the next chapter. Again, the use of this symbol is a matter of convenience. The off page connector is a deviation from the ANSI Standard. It is presently available on the IBM flowchart template.*

OFF PAGE CONNECTOR

TRANSMITTAL TAPE

The transmittal tape is a nonstandard symbol, but is available on the IBM flowchart template. It can depict an output of a column printer (alpha-numeric), adding tape or similar control information (batch, monitoring, etc.). It is document proof, verification, or result of an operation that is not satisfied by the flowchart document symbol. The IBM representation for this operation is shown below.

TRANSMITTAL TAPE

EXERCISE

13. Flowchart a cash register sale in which the purchaser has a credit card and wants a receipt for this purchase. (HINT: Use credit card symbol and transmittal tape — both are nonstandard symbols).

    ANSWER

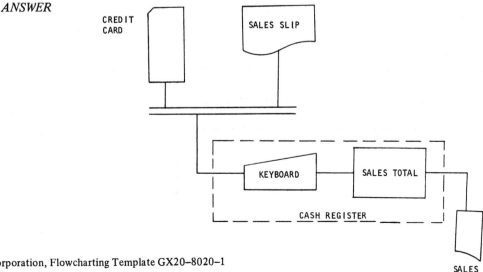

*IBM Corporation, Flowcharting Template GX20-8020-1

SUMMARY

The set of flowchart symbol standards has been defined and illustrated in this chapter with the exception of the branch table and multiplicity of representation. These latter are defined and illustrated in the next chapter.

The vocabulary introduction has been (computer) language independent, although the illustrations are computer oriented with computer components. The exercises and problems are intended to reenforce mental retention and application.

A major problem is the generic and specific application of I/O symbols. When used, they typically define specific formats, data structures, coding, interface configurations, channel assignments, and numerous software requirements. The symbols themselves are not so explicit, but the flowchart user must have this information available if he is to really understand the procedures that follow. Obviously, the flowchart symbol identifies the required information and it behooves the programmer to review this material because it has a direct bearing on data handling and internal memory layout and partitioning. Normally, the I/O data processing is unique for each configuration, since it is the adaptor between the computer and its external environment. The I/O (software and hardware) is the transition area of accommodation for computer entry compatibility.

Digital data processing is becoming very commonplace throughout industry. Each industry has its own vernacular and provincial expressions that it wants to retain despite the inroads of computers. Much of such specialized dialogue can be accommodated with the present set of flowchart symbol standards and conventions. No doubt the present flowchart vocabulary will expand to include the most used expressions.

REFERENCE

1. Rossheim, Robert J. "Report on proposed American Standard Flowchart Symbols for Information Processing." Communication ACM. October 1963, pp. 599–604.

PROBLEMS

1. Define online and offline operations and give examples of applications or system configurations.

    ANSWER

    Not supplied.

2. Define real time, near real time and nonreal time and give examples of system applications.

    ANSWER

    Not supplied.

3. Using four blocks (input, processor, output, console) draw a closed loop control configuration where the console operator is in the loop.

    Answer

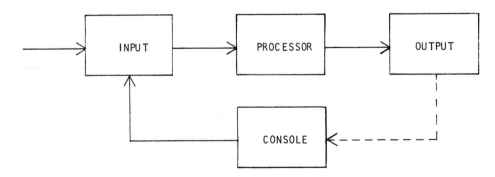

4. Draw the same system as defined in Problem 3 with flow chart symbol substitution.

ANSWER

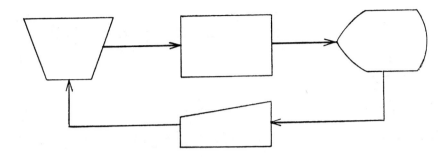

5. Draw a flow chart for problem 4.

ANSWER

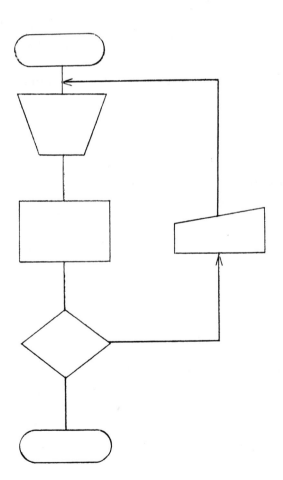

6. Using the decision symbol, flowchart the extraction operation of an unsorted card deck into four suits.

ANSWER

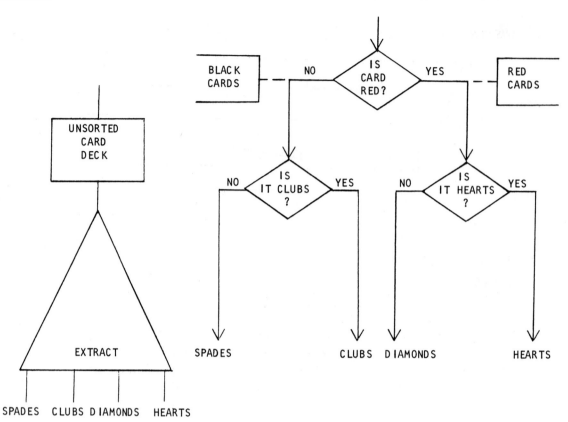

7. Flowchart a disc file update.

ANSWER

Not supplied.

3

FLOWCHART DETAILS and CONVENTIONS

As mentioned in Chapter 1, flowcharts can be an effective means of communication. In order to serve such a purpose, however, flowcharts must conform to standards and conventions of format and usage. The situation is analogous to our English language. Sentence structure, component sequence, phrases, and procedures must conform to grammatical guidelines and conventions which serve as a pliant base for the exercise of lucid exposition and artistic expression. This chapter presents flowchart conventions and techniques.

FLOWCHART FORMAT

In many cases, flowcharts may occupy several pages. The several pages may present either a detailed flow of the system or a top down structure with several levels of complexity. The latter type of presentation is called a "multilevel flowchart." Each level serves a special purpose, detail, or audience. Where there are several levels of complexity and/or numerous flowchart sheets, an indexing system and coding scheme is necessary to identify various systems, subsystems, and parts within a system. The coding arrangement must serve each system's function and assist documentation into a workable set of records for dissemination. The specific coding scheme (alphanumeric), code field length, group arrangement, and level of ascending or descending hierarchy are developed by the user (programmer, installer, manufacturer, etc.) according to his own particular requirements. A suggested flowchart format and organization are presented here to cope with the multilevel flowchart configuration.

Large multilevel flowcharts should be prepared using identification and reference methods already proved by use in drawings of large intricate equipments. The top level chart contains all the pertinent system functions.

Referring to Figure 3.1, note that the top level flowchart correlates all the remaining flowcharts and relevant information to fully detail the system. The flowchart information on the top level chart integrates or interfaces the individual program modules containing the microsoftware details. The magnitude of the recorded information may require sheets larger than the common 8½" x 11" page size. Generally, one of the dimensions is fixed by roll width (18", 24", 30", 36", etc.) and the other is variable. The fixed length contains alphabet symbols for coordinate positioning along this flowchart side. Since the other chart dimension (variable) is not known at the beginning of the flowchart drawing, a decimal coordinate numbering scheme is used. In this manner, all system operations are located at the low numbered coordinates (left side), and all other details and unrelated data are assigned to the right side (i.e., at the higher number coordinate positions). Thus, the top level drawing contains all the information of the system and the breakdown into parts of all details to define the system.

The top level and overall flowchart is shown in Figure 3.1.. A typical format is shown with an expanded legend detail in the lower half. Several methods of connector symbol referencing have been mentioned, and another one is shown here.

When reading a flowchart, one begins reading the flowchart outlines in ascending number sequence. During the first time through, any inconnectors encountered in the flowchart are deliberately ignored. Upon reaching an outconnector (or outconnectors), one must find the associated inconnector to maintain continuity. The tabulation arrangement simplifies this process of locating the inconnector and the next corresponding operation. The outconnector label is searched in the Item Number directory (see Figure 3.1). Once found, it is checked to see that the coordinates are the same. Remember, there can be more than one outconnector for each inconnector, and vice versa. However, the directory is organized on an outconnector basis. After the outconnector is found in the directory list, the corresponding inconnectors are identified as to their locations and next operation. Furthermore, where multiple exits are required, all associated functions (inconnectors) are immediately identified with their corresponding locations.

The reading loop is closed when the inconnector logical operations are connected to the corresponding outconnector. The tabulation of all in and out connectors in one location on the chart reduces the searching procedure in examining system details.

Another frequently used tabulation is the stop/start routine. Generally, there are a number of points, monitoring (printout), test (checking), halts (error), and other branching requirements, including subroutines. The use of connectors for exit and entry points is tabulated to assist in locating each branching point for more detail. The RETURN operation may encompass complex operations including manual intervention. All of these details are centrally located for accessibility. Subroutines, available functions, function generators, and calling operations are also tabulated in this manner. The related information in the far right of the flowchart format complements and augments the value of the flowchart details to its left.

The flowchart form presented in Figure 3.1 acts as a top level overall system description and supplies all the required details and sublevels of information.

THREE-DIMENSIONAL FLOWCHARTS

Flowcharting is a two dimensional graphical presentation. The third dimension is obtained when the level of detail per surface is specified.

A stacking structure is involved (see Figure 3.2). The top layer or level not only serves as a summary, but also serves as the equivalent of a directory of the other levels; it identifies the levels and related items within any one level and among all the levels. Since the top level contains all the basic information in abbreviated form, its projection (down direction) at each succeeding level is an expansion of one or more parts of the previous level. The detailed expansion is continued down the stack until the last level or greatest detail terminates the stack. Here, a one-to-one mapping at any level can be expressed as follows:

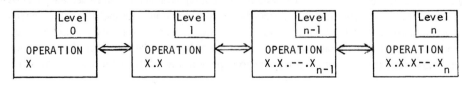

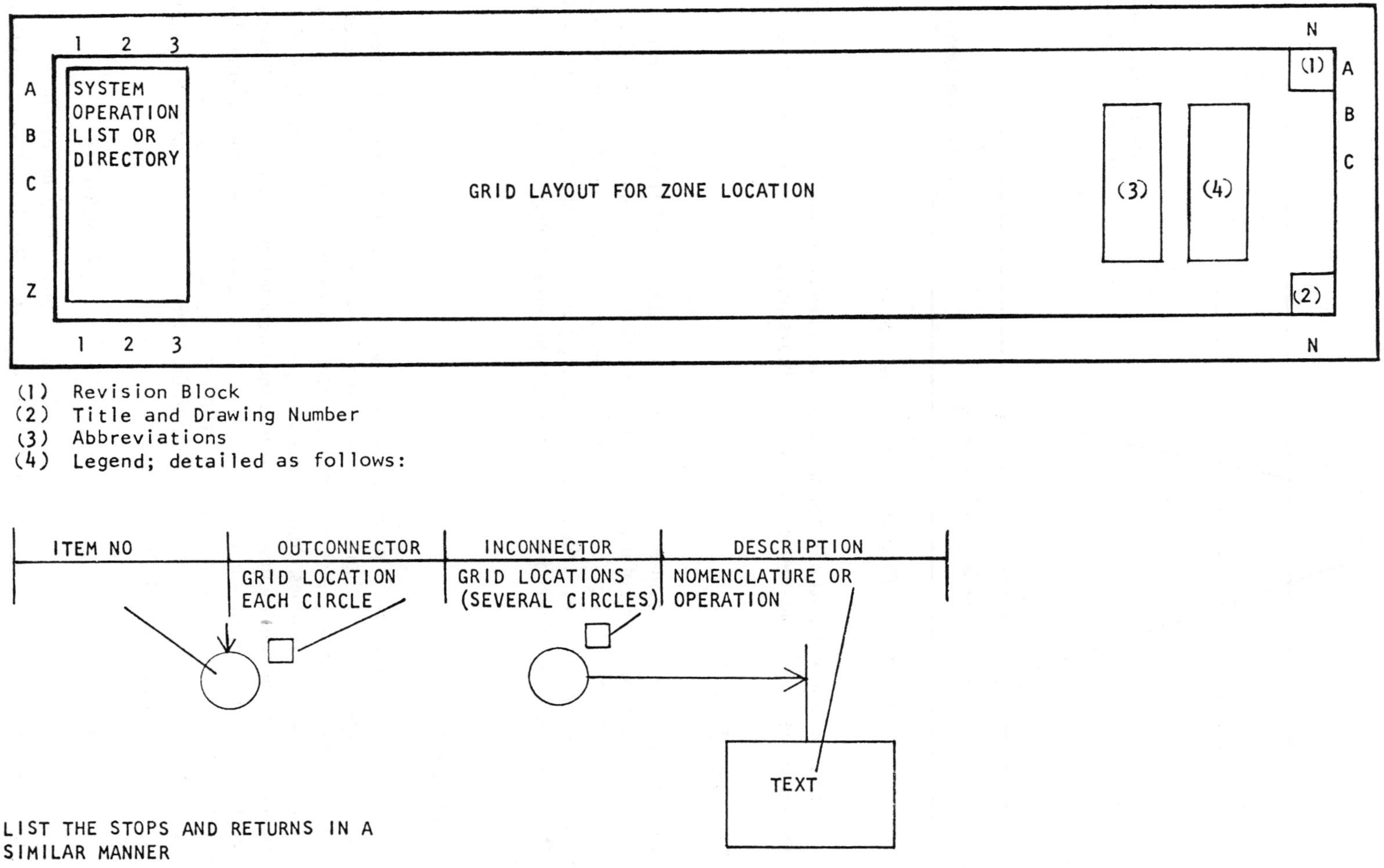
1 2 3
A
B
C
Z
1 2 3
SYSTEM
OPERATION
LIST OR
DIRECTORY
GRID LAYOUT FOR ZONE LOCATION
(3)
(4)
N
(1)
A
B
C
(2)
N
(1) Revision Block
(2) Title and Drawing Number
(3) Abbreviations
(4) Legend; detailed as follows:
ITEM NO
OUTCONNECTOR
GRID LOCATION
EACH CIRCLE
INCONNECTOR
GRID LOCATIONS
(SEVERAL CIRCLES)
DESCRIPTION
NOMENCLATURE OR
OPERATION
TEXT
LIST THE STOPS AND RETURNS IN A
SIMILAR MANNER

FIGURE 3.1 Organizing a Flowchart

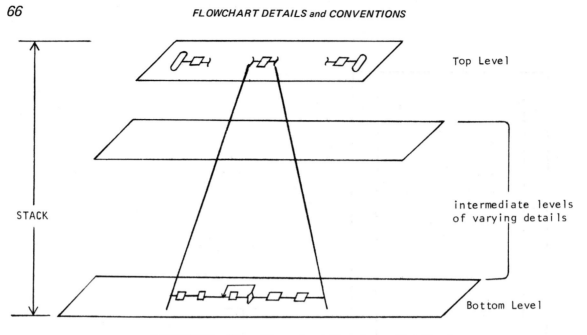

FIGURE 3.2   *Three-Dimensional Flowchart — Multi-level*

For any particular operation of a given level of detail on a flowchart, a higher or lower level of detail can be obtained by cross referencing, and the one-to-one relationship is always maintained as expressed above.

FLOWCHART WORKSHEETS

It is a matter of preference whether a preprinted flowchart symbol layout (Figure 3.3), a coordinate axis sheet (Figure 3.4) or just free form (blank sheets of paper) is used to draw flowchart symbols at the detail level. The 8½" x 11" (or 11½" x 16") sheets are easier to handle than the larger sizes and need not be folded when mounted in binders. The range of flowchart symbols at the detail level (see Chapter 5 for basic primitives and structured program modules) are few, and a two axis coordinate layout system simplifies drawing straight flowlines (vertical and horizontal).

FLOWCHART HIERARCHY

In Chapter 1, the terms "system flowchart," "program flowchart," and "detailed flowchart" were mentioned but not defined. Also, "flow diagram," "logic diagram," and "summary flowchart" are frequently used to define a level of flowchart detail. The flowchart symbols are not necessarily restricted to any level of software design details, whether concept, system, or routines. Contrary, the flowchart content will indicate the level of detail and the audience that is addressed. The construction of a flowchart and its content is not an equation that can be rigorously proven. Instead it is the summation of talent, ability, and knowledge on the subject matter. To this end, this chapter presents existing techniques and conventions widely used and accepted in the field.

Flowcharts can exist at various levels, and there is no limit to the extent of detail that can be flowcharted. Flowcharts have been generally classified into three categories:

CATEGORY	*DESCRIPTOR*
1. Top Level	System, Concept Program, or Summary
2. Intermediate Level (Varying degrees of complexity.)	Program Flow, Procedure Flow, Logic Flow, Logic Diagram, Program Module, etc.
3. Detailed and Final Level	No further sublevels of detail; Software — statement level Hardware — bit level

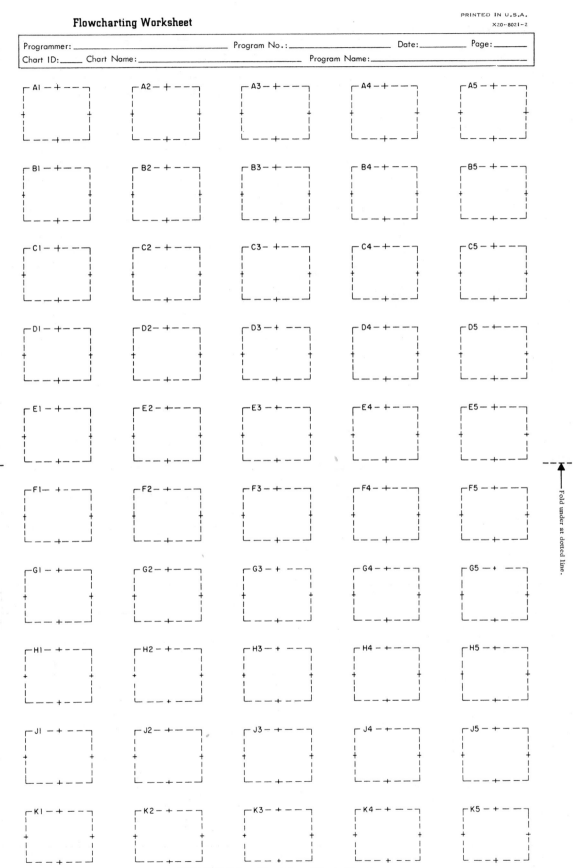

FIGURE 3.3 Flowchart Work Sheet

DIAGRAMMING AND CHARTING WORKSHEET

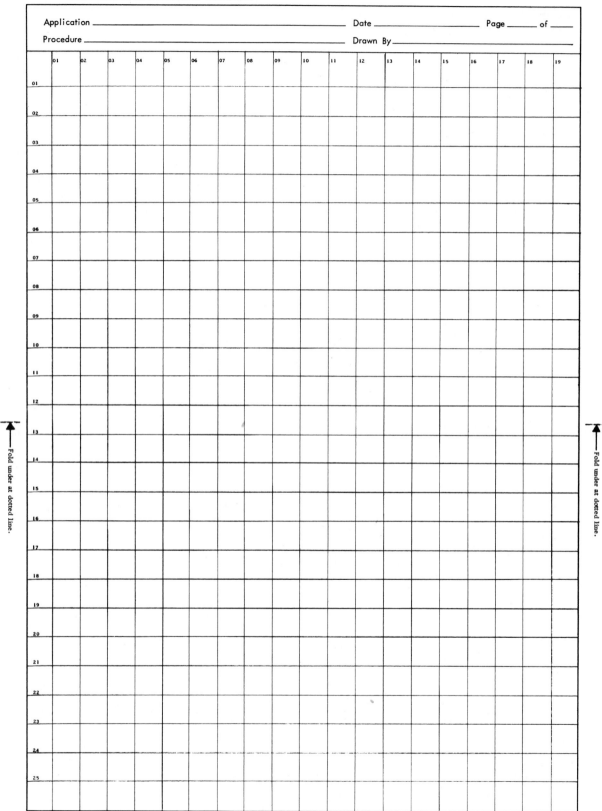

FIGURE 3.4   Flowchart Work Sheet

The level of detail is defined mostly by the user, the system analyst, the programmer, or the software designer. However, the flowchart hierarchy is frequently aligned with the software hierarchy:

1. Total System (Software and Hardware)
2. Subsystem
3. Functions
4. Routines
5. Subroutines
6. Segments
7. Statements

For communication purposes, the three categories are presented as a continuous flowchart flow from top level with succeeding degrees of detail and a final detailed level at the statement level.

The systems flowchart is a representation of the flow of information through the components of a processing system from source document through computer processing to the final disposition of output data. The three basic components of the system flowchart are descriptions of the input data structure and form, the program process, and the output data structure and form. The systems flowchart tells what and how in broad terms. It does not supply the details of the process or the way in which the computer accomplishes a data handling job. Typically, the system uses a greater variety of flowchart symbols (e.g., System and Input/Output flowchart symbols). The system flowchart is used to block out the system and highlight a main process.

Generally the system flowchart begins with input data identification and always ends with output data identification. As a rule, but not always, the terminal symbols are not present, because the I/O symbols perform the start and stop operations. A simplified system flowchart for a payroll operation is shown here.

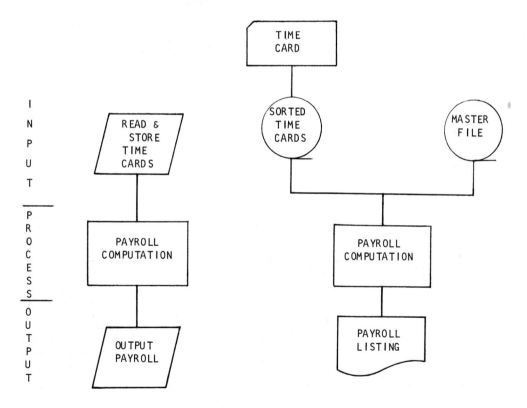

Note that both flowcharts are adequate in detail. The flowchart on the right is more descriptive and details the input sources, media, and the output document form. The basic operation is a payroll computation, and that alone is adequate information to enable one to assemble and schedule the resources at the data center facilities. On file will be the time cards, employee master tape file, I/O routines, and the computer program tape (not shown).

Now the computer process will be further detailed to supply program details. The system flowchart is detailed to show how the time card data is used to make a payroll computation (see Figure 3.5). The time cards are sorted prior to payroll computation. The payroll computation for a single time card is added to the flowchart.* The next level of detail expansion repeats the process for the next time card payroll computation.** At this point, data flow is presented as a sequential process and the computer operation has not been included. The next flowchart level includes the details to complete an automatic data processing operation. The program flowchart (or logic diagram) helps identify the logical steps necessary to solve a particular problem. The program flowchart differs from the system flowchart in that the system flowchart indicates the hardware, identifies the files, and shows the general flow of data; whereas the program flowchart defines the detailed logical steps involved in the input, output, and processing of information.

There are several observations worth mentioning concerning the program flowchart in Figure 3.5. The input and output terminal symbols are not shown and neither are the details of hardware or equipment.

In most cases the system details can be flowcharted on a single page. Not so with most program flowcharts. When necessary, the program flowchart is broken into parts and segments, requiring connector usage and cross referencing.

The techniques of modularizing, segmenting, and partitioning are presented later. A very common feature of procedure implementation using the computer is the ability to repeat (cycle or iterate) an operation. This appears as a loop as shown in Figure 3.5.*** In fact there are two loops in series and another loop that encircles these two (e.g., nested). And finally, there are a limited number of flowchart symbols at the detailed level. The basic symbols together with the additional symbols in Chapter 2 are sufficient to present all the programming details of algorithms for problem solving.

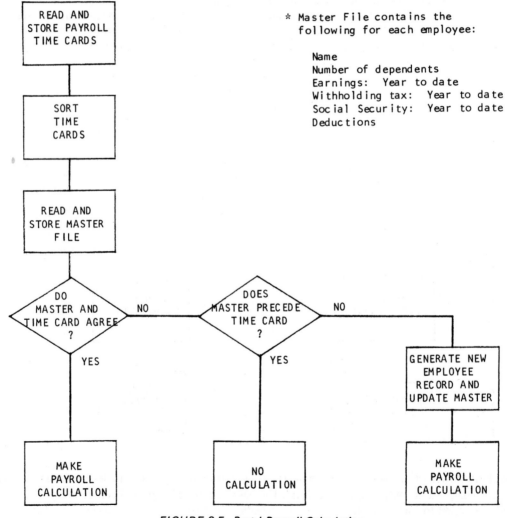

FIGURE 3.5 Part I Payroll Calculation

*Part 1 of Figure 3.5
**Part 2 of Figure 3.5
***Part 3 of Figure 3.5

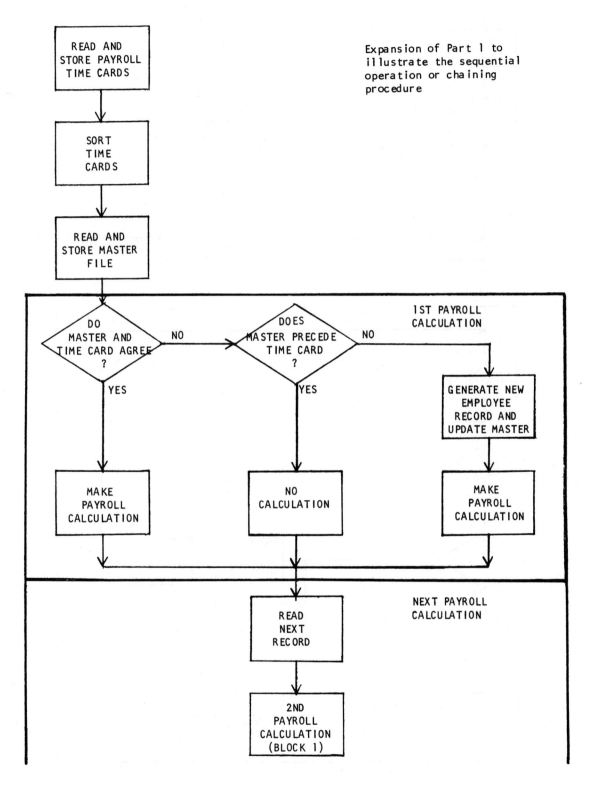

FIGURE 3.5 Part II Payroll Calculation

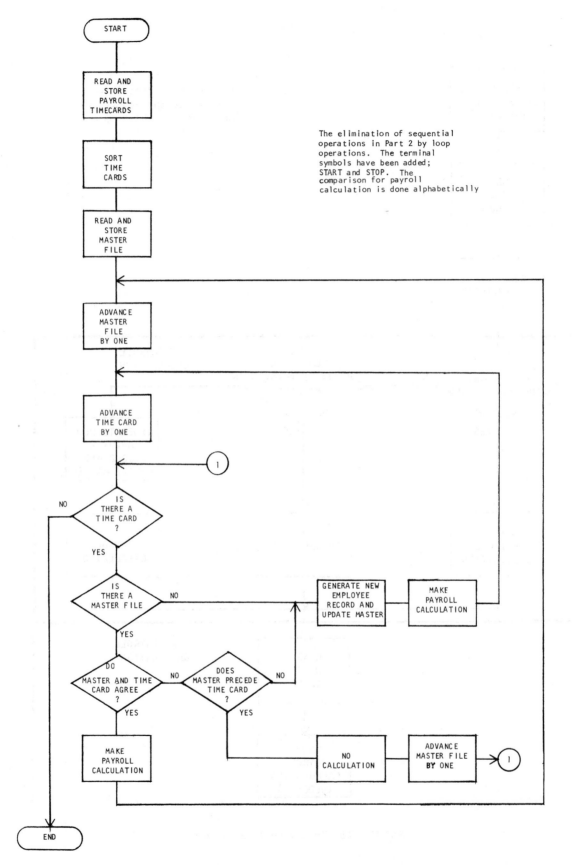

The elimination of sequential
operations in Part 2 by loop
operations. The terminal
symbols have been added;
START and STOP. The
comparison for payroll
calculation is done alphabetically

FIGURE 3.5 Part III Payroll Calculation

The program flowchart shown in Figure 3.5* is rather detailed in comparison with the system flowchart representative of the payroll computation. Yet it is still quite brief or concise when compared to the statement level. A typical flowcharted statement is shown here. Obviously the chart in Figure 3.5 Part 3 depicts far less detail than that which the computer program needs.

FORTRAN

SUM(1) = HOURS*BASRATE

The reader may think that the program flowchart (Part 3 of Figure 3.5) should be the top level presentation, since it is only slightly more detailed than Parts 1 and 2 of Figure 3.5. Now the audience must be considered. For the data processing manager, the system flowchart is adequate for his needs. To the system analyst or software system designer, the program flowchart will suffice. If coding is considered, the program flowchart must be further detailed, but not necessarily down to the statement level. Flowcharts for coding need not be drawn at the statement level. The final level should contain sufficient information to permit coding whose logic accomplishes the intended payroll computation as defined in the program flowchart.

Although the flowchart vocabulary is now available, and a selection of flowchart level of detail or audience to be addressed is made, there still remains the presentation techniques to create a flowchart. Drawing a flowchart is an art. The remaining portion of this chapter is devoted to developing the artistic talent of the reader.

EXERCISE

1. Complete the third column for on line storage.

ANSWER

1ST LEVEL | 2ND LEVEL | 3RD LEVEL

INPUT/OUTPUT | ONLINE STORAGE

ANSWER – SYMBOLS
1. *Magnetic Drum*
2. *Magnetic Disc*
3. *Magnetic Tape*
4. *Core*

2. Complete the third column for off line storage.

ANSWER

1ST LEVEL | 2ND LEVEL | 3RD LEVEL

INPUT/OUTPUT | OFFLINE

ANSWER – SYMBOLS
1. *Magnetic Tape*
2. *Punched Tape*
3. *Punched Card*
4. *Document (via the keyboard)*

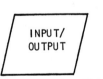

*Part 3 of Figure 3.5

3. Given computer inputs as alpha–numeric display console (keyboard), discs, magnetic tapes and paper tape, draw an appropriate computer loading operation for an input system flowchart.

ANSWER

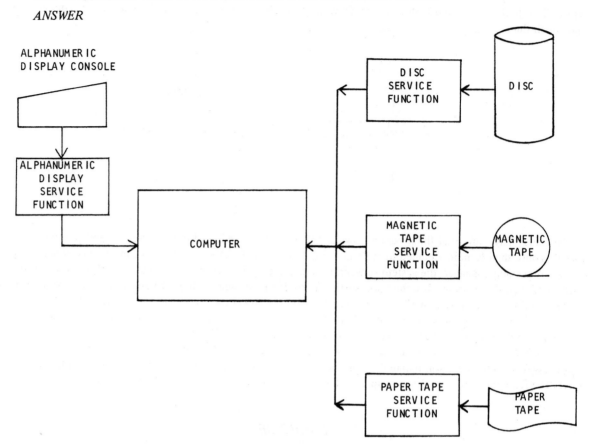

4. A top level manual operation is depicted by the following symbol.

OFFLINE

Flowchart the next level for printed material for computer entry

ANSWER

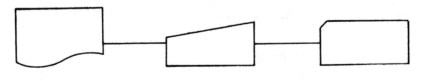

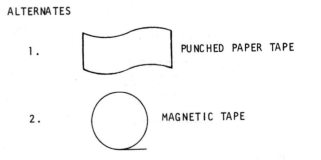

ALTERNATES

1.   PUNCHED PAPER TAPE

2.   MAGNETIC TAPE

5. Draw a system flowchart for the selection of the largest of three numbers A, B, and C.

    ANSWER

6. Draw a program flowchart for the selection of the largest of three numbers A, B, C in sequential (series) form.

    ANSWER

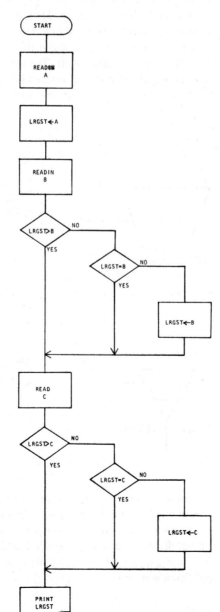

PRESENTATION TECHNIQUES

The visual impression of the geometric configuration enhances memory association, retention, and application. Also, slight imperfections or distortions of the flowchart symbols do not convert nor transform them into other valid symbols for possible misinterpretation. Uniqueness is obtained from the symbol selection so that its contents and outline denote the basic information. Auxiliary notations, text material, coding, striping, branching, coordinate referencing, and tabulating complement the flowchart symbols and enhance this language as a dialogue medium.

SYMBOL ORIENTATION

The orientation of each symbol on a flowchart should be the same as shown in the preceding chapter. Therefore, regardless of the flow direction represented, horizontal and vertical, they should remain the same as they were presented. Generally, a majority of the symbols possess a symmetry about the horizontal and vertical axes. Also, there are a few that have directional properties such as the display, manual input and operation, on-line and off-line storage symbols. Most frequently, due to ease of drawing and handling automated printouts and so forth, the direction flow is from top to bottom and continued left to right until a page is completed using connectors to maintain flow continuity from bottom to top and column structure. There are many symbols having both vertical and horizontal symmetry. The symbols with no symmetry pose no orientation problem for normal direction flow. In some cases, the display and on-line storage symbols are rotated about the vertical axis to take advantage of the directional property in the flow direction left to right or right to left. The manual operation and off-line storage symbols favor the top to bottom direction. However, they are not rotated (90 degrees) for horizontal flow direction. Instead, these two symbols are shown for directional flow left to right in the following manner:

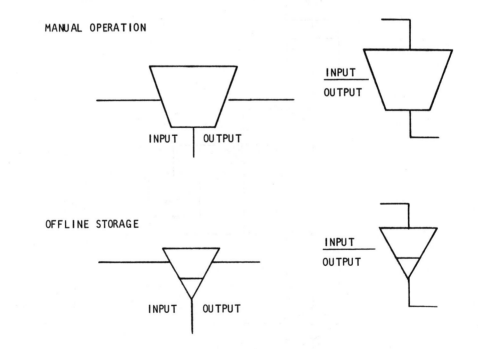

As stated earlier, all symbols that perform an operation or process information and data have a minimum of one input and one output, excluding those symbols that are control and editing such as the connector, annotation, flowline, etc. Only the decision symbol requires a minimum of two outputs. Therefore, a line perpendicular to direction flow for entry and exit of a symbol divides or bisects the symbol artificially into an input and output section. The designated sections denoting input and output are applicable to all symbols and not limited to the two previous illustrations. This additional detail is necessary to expand the scope of flowcharting to include digital hardware and concurrency.

FLOW DIRECTION

Flow direction is represented by lines drawn between symbols. Repeating, normal direction flow is from left to right or top to bottom (see Figure 3.6). When the flow direction is not left to right or top to bottom, open arrows (←) are placed on reverse direction flowlines. When increased clarity is desired, open arrowheads can be placed on normal direction flowlines. Slant, curved, and diagonal lines should be avoided. When flowlines are broken, connector symbols are used to indicate the break (see Figure 3.7).

When flow is bidirectional, open arrowheads are used to indicate both normal direction flow and reverse direction flow. Since there is an input and output per symbol, it is preferable to maintain this distinction rather than use a single or double line with open arrows at each end.

MAIN BRANCH FLOWLINE — MINOR BRANCH

If the flowchart (or computer program) were just a series of symbols (no branches), the need for flowcharting would not exist and automatic data processing would be severely limited in scope of operation. However, this is not the case.

The use of a decision symbol requires two (or more) output paths. One of the decision symbol outputs should be chosen to emphasize what the programmer has in mind to maintain the (main) flow in the vertical direction (top bottom flow is considered here). The remaining output(s) is from the left or right side. For each encountered decision symbol, one output is designated a major flow in the branch and the remaining output is assigned a minor role. Obviously, there is the alternate patterns of main (major) flow and branches (minor). An example of an obvious flow path is shown in the diagram.

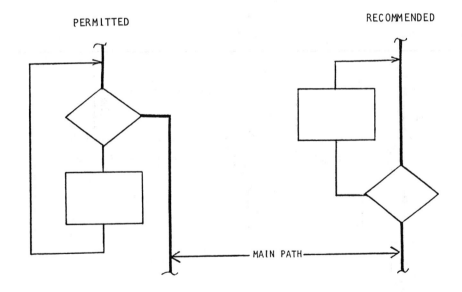

The main flow path concept is not a trivial one, for it has serious consequences if not chosen correctly. The main path may be the most frequently exercised path, and the remaining flow paths are supporting and auxiliary operations. By definition, a branch path is a conditional consideration or deviation from the main path that the programmer may want to de-emphasize.

For testing purposes, the main path is frequently chosen as the preferred or first thread to be tested as a chain in order to exercise the highest percentage of statements in a module, or the highest percentage of modules in a program. Under certain circumstances, however, the reverse could be true. The shortest path might be chosen to simplify a test and to minimize the number of links or stubs to be used in the test. The main flow could reflect the availability of modules to accomplish a complete chain for testing or an inked path can designate the same.

The questions being posed can only be answered by the programmer. If a number of persons were to read and identify the main flow of a computer program, the main flow should occupy one column (a single branch). If this is not possible, the main flow (main branch or major chain) can be emphasized by a heavy flowline joining each flowchart symbol to another symbol so that it is obvious to the reader.

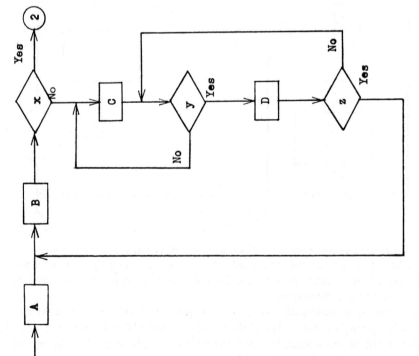

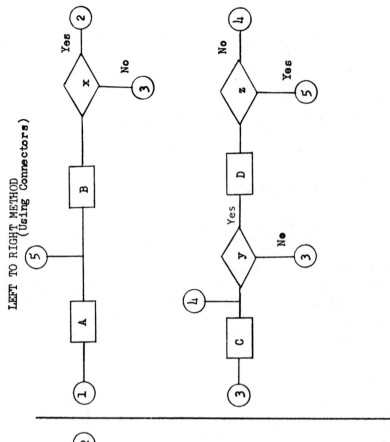

FIGURE 3.6　Flowchart Style

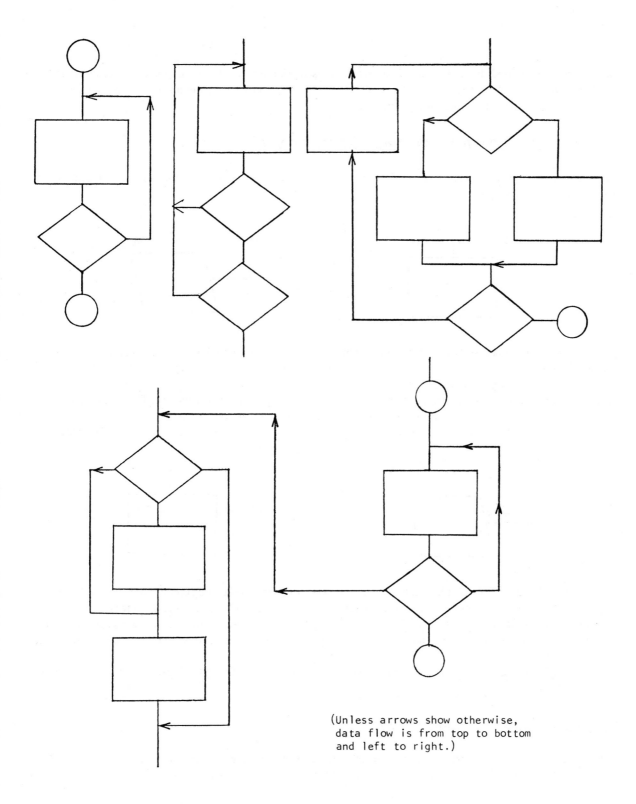

(Unless arrows show otherwise,
data flow is from top to bottom
and left to right.)

FIGURE 3.7   Flow Direction Conventions in Use

SYMBOL SIZE

The size of each symbol may vary, to accommodate the encircled text, but the symbol outline must be retained. Size and location of a symbol on a diagram could also serve to indicate its relative importance in the process or routine being flowcharted. In order to obtain an artistic presentation, the symbol ratio dimensions as specified by the X3.5 ANSI Standard has been waived. Generally, the use of a flowchart template insures conformance to X3.5 ANSI Standard, but this is not always the case.

SYMBOL IDENTIFICATION AND CROSS REFERENCING

When identifying a symbol to facilitate its reference to other elements for documentation, a label is placed on the right side and above the symbol.

Generally, all that is necessary is a numbering scheme. Remember, the symbol location and identification is obtained from this number while the page number and title are the next level for referencing. The symbols may be numbered from 1 up, beginning with the first symbol of the flowchart of the system. A different number is assigned to each operation in the flowchart and in normal operation performance sequence. The symbol number (or box number) is the basic indexing tool for maintaining an orderly set of operations.

When cross referencing a symbol to a line of coding, the appropriate identifier is placed on the left side and above the symbol.

Typically, the cross referencing is to source card numbers. No doubt the labeling content depends on the overall system documentation. Therefore, the cross referencing must be compatible in a programming system or a particular system using data sheets, printed forms, record layouts, card decks, or even media such as paper tapes, magnetic tape and discs. In any event, cross referencing is provided in both directions from flowcharts to reference sheets and vice versa.

The "to and from" and "from and to" approach helps avoid errors in defining the location of the data sources in the system.

CONNECTOR SYMBOL REFERENCING

Quite often a complete flowchart will require several sheets of paper. Each person may have his own method of maintaining continuity of thought, even to glueing the sheets together to form a scroll.

The connector symbol was introduced earlier to satisfy the problem that arises where a flowline is broken by any limitation of the flowchart. When two connectors are required to maintain flowline continuity on a single page, the reference problem is resolved by a number or label identification within the symbol.

The method of connector referencing between pages and among other documents conforms to the foregoing convention. The connector symbol is assigned a label as an outconnector (source or originator) or an inconnector (receiver or recipient). The outconnector is linked to its associated inconnector by means of a connector identifier placed above and to the left of each connector. Reference to the inconnector is shown by including the chart page (or other reference notation) on which (or where) the inconnector is found. The reference to the outconnector from the inconnector is shown by listing outside and to the left of the inconnector the chart page on which the associated outconnectors are found.

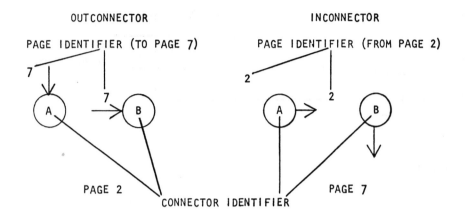

As shown, the inconnector is referenced from the outconnector (from rather than to). Also note the flowlines terminate at the outconnector.

SYMBOL STRIPING

Symbol striping is used to indicate that a more detailed representation of a function is to be found elsewhere on the flowchart or cross reference. A horizontal line is drawn near the top of the symbol, and the identifier can include the chart number or page followed by the inconnector identifier.

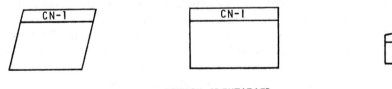

COMMON IDENTIFIER

CN = Chart Number or Page

I = Identifier or Symbol Number

An outconnector is drawn to show the return from the detailed representation to an inconnector or the symbol immediately following the striped symbol.

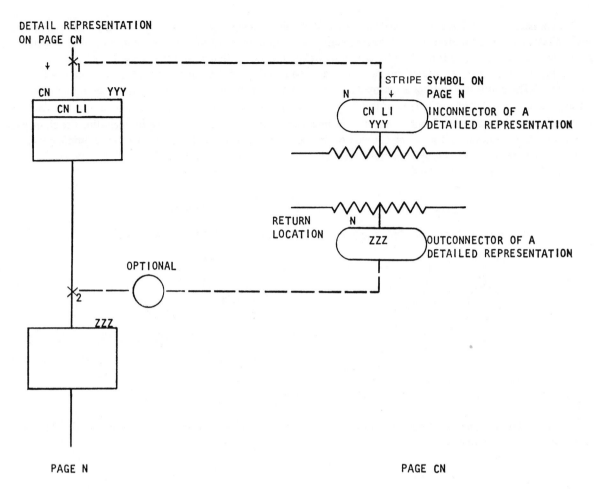

In the example, the symbol YYY has a more detailed representation on page CN starting with the inconnector CN-LI, YYY. It is understood on page CN that the complete details of YYY are present (or segregated) so that the set, inconnector and outconnector, correspond to X_1 and X_2 respectively. Actually, the outconnector has a pseudo function. It is necessary to note the boundary or terminals of the functions of YYY. It is taken for granted, that after completing the details on page CN, the normal return is immediately to ZZZ. In recalling the preceding section on code referencing, the written material in the strip portion may not be as simple as shown, but may comprise text information that is continued or expanded in the main section of the box. The inconnector and outconnector symbols are used to denote the extremities of a detailed representation description. The dashed lines used here are for illustrative purposes only. A current technique using connector symbols that is in common practice is as shown in the diagram.

The use of connectors is equivalent to the striping convention of detailing the representation of a function elsewhere on the flowchart.

The annotation symbol could have been used as a means of identifying striping substitution. Also, the detailed operations or program steps may comprise a subroutine or logical unit of information that can be called a predefined process.

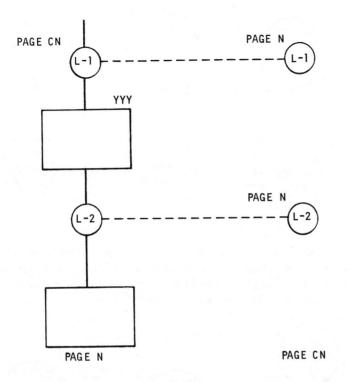

A predefined process that is acknowledged and accepted as a routine would generally be recorded with a striping on a predefined process symbol. However, this method is no longer acceptable and is in violation if the detail representation is not part of the flowchart material.

The striping convention is a compact method of identifying a detailed function most efficaciously.

CONVENTION TECHNIQUES (MULTIPLE METHODS)

Up to this point, the discussion has been limited to linear and singular conventional methods of flowchart representation. The graphical presentation has been limited to a two dimensional or flat surface with consideration to the prime use and value in organizing, planning, and comprehending a complex system and its procedures. Multiple techniques add another dimension to flowcharting with considerable complexity. A number of methods have been used in the past prior to this technology and a few of the accepted conventions of multiplicity of the same item or a group of items are reviewed here.

MULTIPLICITY REPRESENTATION

Flowchart symbols may be shown in an overlay pattern to illustrate the use or creation of operation multiplicity. A typical multiple media, devices, or files are shown in the diagram for a digital magnetic tape recording device.

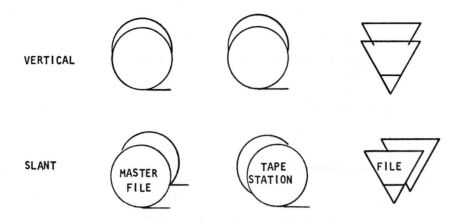

A choice of vertical stacking or slant assembly is shown here. The number of repetitive symbol drawings is a matter of choice and information content. Suppose if the multiplicity were five or less, would you draw five symbols? Possibly. Suppose there are 10 items, functions or operations — how would they be drawn? Some of the techniques are shown in the diagram. Here, the same item is repeated 10 times.

There are numerous examples where overlay patterns to be illustrated have some unifying or labeling cohesiveness, but each symbol is unique in its contribution to a set, block, or group of operations that appear to be a stream or series chain.

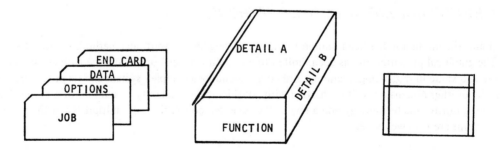

In the illustrations, the group of operations have a unifying label or identification. They are represented by an overlay, a three dimensional block and a predefined process symbol with striping disclosing the necessary detail at another location.

The introduction of a three dimensional block presents three different visible planes with coordinates that can be correlated or the front plane contains the function of the block that is customary for flowcharting and the two other planes may represent information yet to be specified.

This multidimensional method retains the sequential relation and physical meaning in a relationship to flow-chart drawing. As a matter of fact, the three dimensional blocks can be drawn without difficulty from available flowchart templates. The introduction of three dimensions here is one of suggestion and experimentation.

For simplicity, the complexity can be circumvented by sequential symbol steps. The overlay and three dimension are expressive, dramatic, and should be reserved for these effects, rather than the mundane detail operations for technical interchange.

The ANSI Standard sets forth a priority or sequential order for multiple symbol presentation. Regardless of the symbol outline, the front (first) to back (last) configuration flowlines are input and output respectively. With this in mind, the multiple symbol should be drawn to convey the horizontal and vertical direction information flow as designated in the diagram.

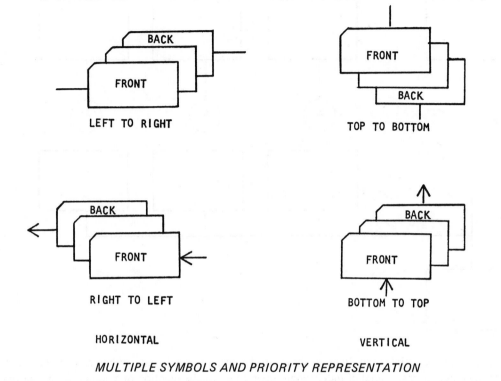

MULTIPLE SYMBOLS AND PRIORITY REPRESENTATION

MULTIPLE EXITS AND ENTRIES

Multiple exits from a symbol and multiple entries to a symbol are very much related. The multiple entry will be examined initially.

By definition, multiple entries are the recombination of alternate paths that converge to a common flowchart segment.

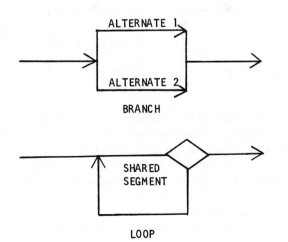

Up to this point, each symbol is joined to another using a flowline for sequential steps. Also, each symbol by definition has only one entry.

For multiple entry operations, the conventional techniques are applicable.

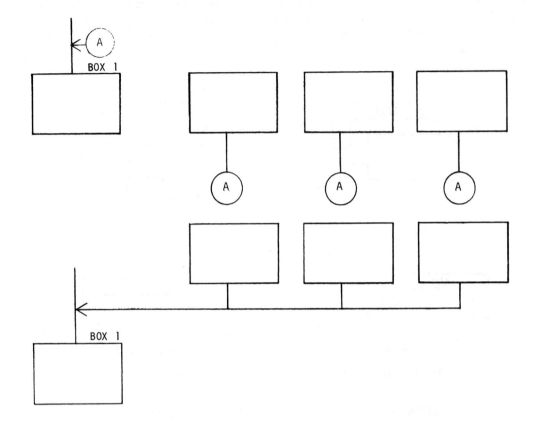

For the moment, the multiple input lines, as shown in the diagram, can be interpreted as several lines of identical or similar operations, several of independent operations, or the junction of a common singular operation.

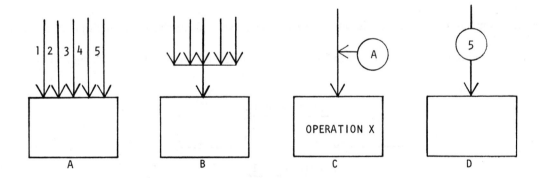

Before we proceed further, note that Method D is valid in other styles of diagramming and drawings (logic), but not here. Methods B and C are identical, with C being preferred. Method A discloses five inputs, but gives no additional information. Generally speaking, flowcharting is a linear and sequential process per symbol. Therefore, each box performs a unique operation and one input line is adequate for entry for the process. Since no stipulated conditions have been stated as yet, Method A is then similar to Method B and should be drawn as illustrated in C. Once Method C is invoked, there is no way of identifying the number of inputs. The information is retained by cross referencing or by tabulation on the flowchart itself.

Method C satisfies the single entry per symbol and the terminal of a multiple flowline junction point. It can be stated that Connector A is in reality Operation X, and entrance into Operation X is Connector A. The expression is drawn in the following manner and contrasted with Method C.

METHOD C

At this point, it becomes a matter of choice, personal style, and taste. Multiple exits from a symbol may be shown by several flowlines from the symbol to other symbols or by a single flowline from the symbol, which then branches into an appropriate number of flowlines.

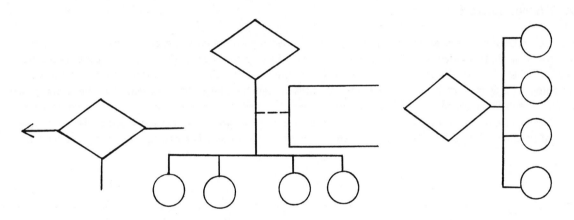

The decision symbol by definition has a multiple output, with a minimum of two required; and as shown, there are numerous possibilities. The identification of each exit, and additional details using the annotation symbol, is permitted by the recording of all of the qualified conditions associated with each individual exit. Numerous exits are available with this method of implementation.

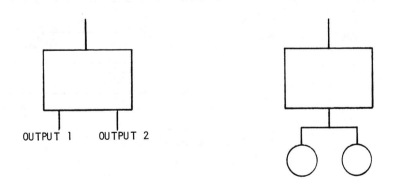

What has been said previously concerning multiple inputs will be repeated here. Each box performs a unique operation, with one input line and one output being adequate for one processing operation. The splitting of a flowline into alternate paths, using connector symbols, states that the common output has two alternate paths so identified (right side drawing). The drawing on the left shows two outputs 1 and 2. This method, in association with the labeling of the symbol, indicates that the drawing is describing not just two outputs, but two unique (differentiated) outputs. A typical example of this arrangement is as illustrated.

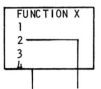

In performing Function X to obtain an output, some intermediate step may be relevant to another operation or may serve as a required input for another processing operation. It may not be advisable or possible, as related to Function X, to draw two boxes in series identifying these two operations. As a consequence, where a single operation is a supplier of several outputs, and the operation is compact or integrated as a unit, then several identified outputs, for a single symbol, may be warranted and appropriate. A typical example where a second flowline output might be used would be to indicate an arithmetic overflow or a divide remainder. For example, the trigonometric parameter theta(θ) is a supplier of sine θ and cosine θ. And similarly, a squaring operation may supply the square root of a number in a closed loop checking routine or hard wired operation.

BRANCHING TABLE

Branching is related to a transfer operation instead of the normal path operation. A branch point is where the branch takes place as specified in an instruction or other qualifiying conditions. Therefore, branching is another form of splitting a line of flow into two or more paths, on the basis of a fixed set of specified values for a single variable item of information. In lieu of connector symbols, a branching table may be used. The table is composed of a statement of the decision to be made, a list of the conditions which can occur, and the path to be followed for each stated condition. A suggested branching format is shown in Figure 3.8 oriented in either the horizontal or vertical position. Within the table, the specified information is recorded. The exiting path values are generally

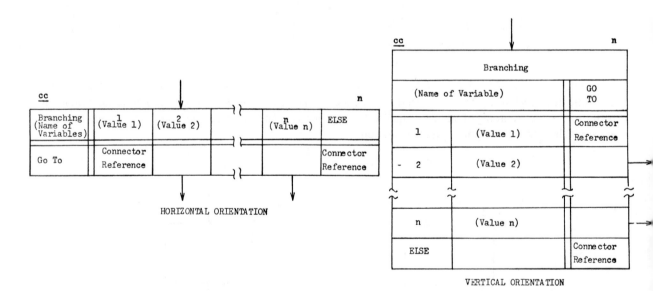

FIGURE 3.8   Branching Table

arranged according to the expected degree of frequency or activity. The most frequently used value is placed in path 1 or the normal flow path. Each "Go to" section contains either a connector reference or a single flowline exiting to another flowchart symbol. Only one flowline is drawn entering this table at the top center, and only one exiting flowline is drawn from the center of each appropriate go to box. When all the variable values have been accounted for each specific condition, all remaining possible values or occurrences can be grouped together and assigned a single exit representing the "I don't care" case or "Else" (see Figure 3.8). The branching table can serve as a flowchart symbol with one input and as many exits as desired.

SWITCHING OPERATION

Switching in a programming routine is quite similar to switching in a logical sense. Based upon the prevailing data and assigned conditions, two or more paths are available on an individual basis or in any combination as specified. These paths are preselected through qualifications and prior modifications of the switch setting based on the requirements of the processing logic. In a typical multiple exit arrangement, one connector symbol serves to label each flowline prior to its splitting into alternate paths.

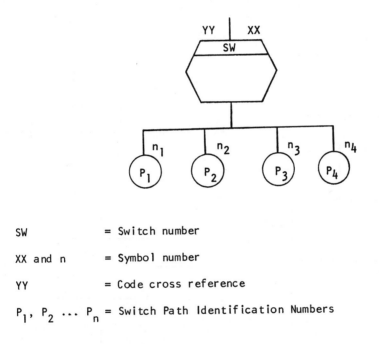

SW = Switch number

XX and n = Symbol number

YY = Code cross reference

$P_1, P_2 \ldots P_n$ = Switch Path Identification Numbers

   For switching, the switch number identifier (SW) may be in the stripe area of the symbol along with the appropriate labeling and code cross referencing information. Alternately, the annotation symbol can be used to explain the need or purpose of each path.
   Multiple exits were previously presented using a decision symbol having numerous paths for the same information. This decision symbol notation permits the tabulation and identification of all recipients of the same information. The branching table and switching operation supply the same information, consolidating and compressing conventional techniques which identify and designate alternate paths that are unique in themselves by representing specific conditions and qualifications.

BETTER ART WORK

Flowcharts are not simple and easily created. First attempts to generate a flowchart may prove to be discouraging, difficult, and even self-defeating.

Flowcharts do not come into being by any magic formula. They start somewhere, possibly with a rough sketch or a layout, and they are taken through successive stages of addition and correction. Gradually, the chart takes form and progress is acknowledged. Final touches are made, and after the clean up process (pencil smudges, extra lines, and erasures), the flowchart may be admired as a work of art. Flowchart drawing is not like the writing of an equation in which parameters (or values) are substituted that result in correct answers. Two charts of the same process may both be correct, but one may be more elegant, concise, and easily understood.

There are always a number of ways of expressing an idea, and flowcharts will show this. The published literature is an enormous source of ideas to emulate and to avoid. A few examples are given here. In most cases, the particular point stressed entails a few flowchart constructs. A discussion follows showing the flowchart construct and a preferred way. A very common hybrid flowchart construct is the data flowline entering and leaving the symbol. The symbol has an input section and an output section that should be observed. If this is considered there is no problem. (In hardware, the input and output notations are enforced.) A typical example is shown in the diagrams.

ADEQUATE

PREFERRED

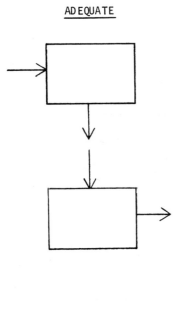
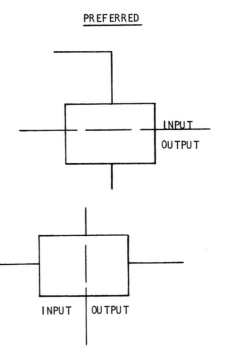

There is always the possibility of misinterpretation with the hybrid (adequate) construct. Arrowheads are necessary. The preferred way needs no further explanation, especially if it has multiple inputs, multiple outputs, or both. Besides, it looks better.

A common error using the connector symbol is frequently seen.

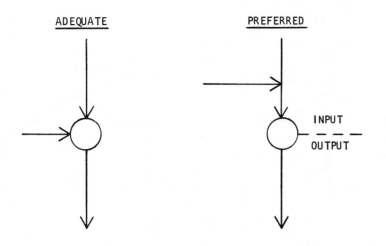

In the literature, the hybrid notation might be confused with the logical operation ANDing and ORing, or possibly with summing (analog technology). These possibilities are drawn here; note that use of the connector symbol in the "preferred" manner eliminates any possible confusion.

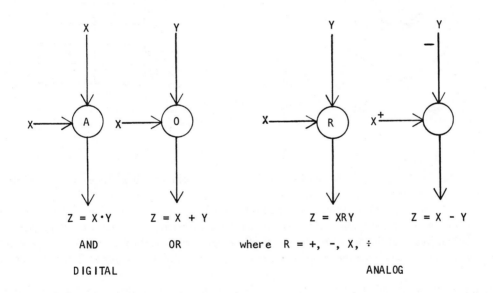

The decision symbol has four points, and some individuals use these points in a number of ways. If more than four inputs or outputs are needed, flowlines are drawn in any fashion for entries and exits. The decision symbol has a single input and has a maximum of three outputs.

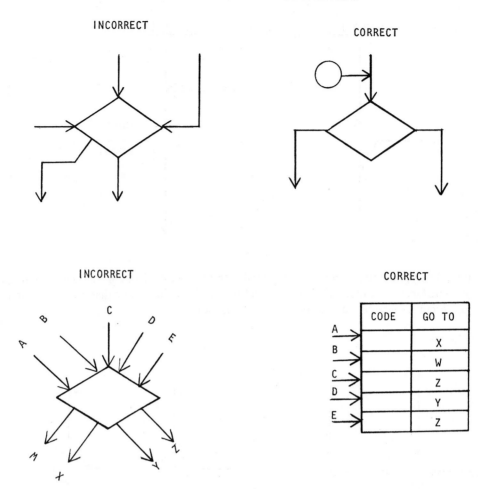

In many cases the decision symbol is overworked and abused. The branch table is an excellent symbol to handle multiple inputs and outputs in any combination.

The flowchart examples in Figure 3.9 were obtained respectively from a programming book (top left) and a paper on a complex computer operation (bottom left). The top left flowchart is redrawn (at top right) to shown the main path or branch and at the same time to give the flowchart a more vertical look. Similarly, the bottom left flowchart is redrawn to eliminate the use of a connector symbol without resorting to flowlines crossing one another.

Figure 3.10 presents some basic recommended methods and conventions for flowchart drawing. The three groupings, preferred, adequate, and incorrect, show what to expect in the literature. The "adequate" column discloses an extension or modification of the present ANSI standards. Some of these constructs may be overlooked, but the methods and conventions in the "incorrect" column should always be avoided. Figure 3.11 shows a vertical and horizontal orientation for labeling the process symbol as a group classification.

There are preferred conventions in using the connector symbol (ANSI) and off page connector (IBM). When representing an outconnector, the symbol has an input and should be drawn to the right side of a flowline or below it, as shown (page 96).

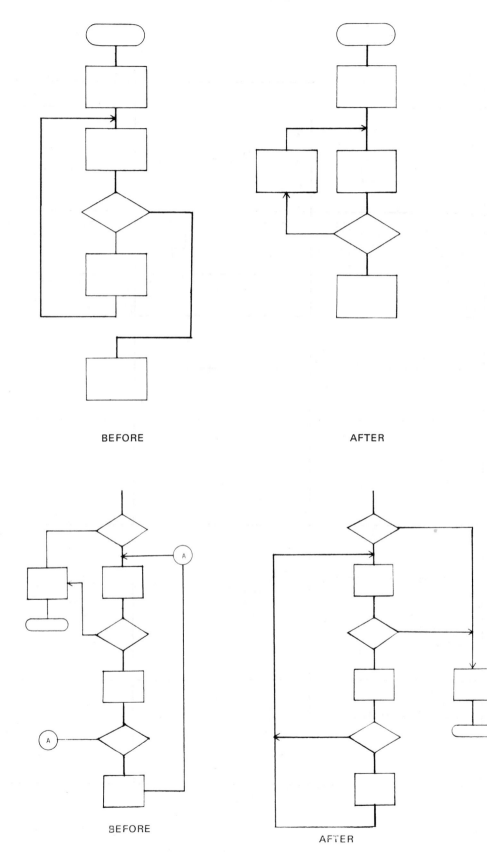

BEFORE   AFTER

BEFORE   AFTER

FIGURE 3.9   Flowchart Main Path Emphasis

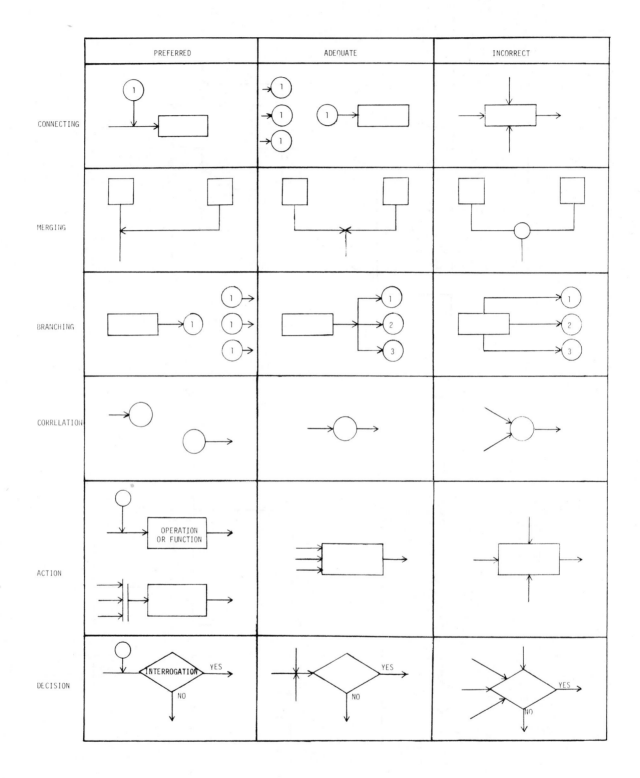

FIGURE 3.10   Recommended Conventions

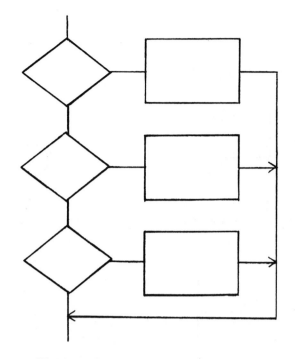

VERTICAL PROCESSING FORMAT

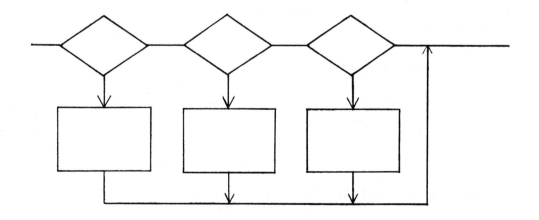

HORIZONTAL PROCESSING FORMAT

FIGURE 3.11   Flowchart Vertical and Horizontal Orientation

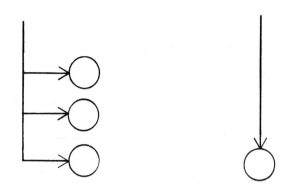

The inconnector symbol has an output and should be drawn to the left side of the flowline or above it, as shown.

If the off page connector is used, then the inconnector and outconnector are restricted to connections on the same page. The off page connector to another page is drawn pointing to the right or drawn at the bottom of the page, as indicated.

The off page inconnector from another page is drawn pointing to the right and is positioned at the left or at the top of the page, as described.

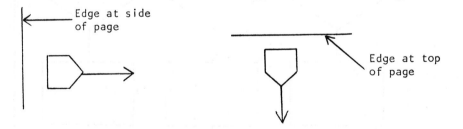

Figure 3.12 shows a common error in many flowchart drawings. Either by carelessness, reworking, or even poor design, unnecessary multiple loops may result. The parallel or separate channel method shown here compresses the information and may disclose an infinite loop.

The reason the flowcharts are drawn in the vertical direction (top to bottom) is a simple one. The vertical page length is the larger length. More symbols can be drawn with a minimum use of connectors or a minimum use of

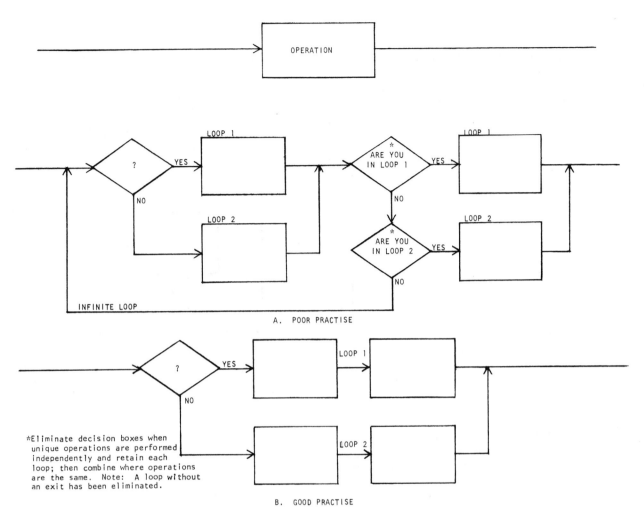

FIGURE 3.12   Flowcharting Independent Branches

flowline length. When the page is oriented with the longer length in the horizontal direction, the flow direction is left to right.

Descriptive titles or text contained within symbols should be brief, but not confusing. For better understanding, the language used in a flowchart is English. Where coding is involved, then high level language and machine oriented language (Assembly) are appropriate and preferred. Wording inside a symbol should be condensed to fit without overcrowding. To avoid ambiguity, any abbreviations or mathematical, logical and relationship notations should be defined prior to usage or consolidated in a glossary.

Two basic methods for making annotations and comments are available. The annotation symbol can be used for infrequent comments. When considerable annotations and comments are necessary, the usefulness of the flowcharts is enhanced by the technique of drawing the flowchart symbols down one half of a page and a corresponding explanatory narrative down the other half. The result is a symbolic running commentary. The commentary is located opposite the symbol flow without the use of the annotation symbol (see Figure 1.1).

In some cases, a numbering scheme is used to associate comments and corresponding flowchart logic symbols.

The best way to analyze any problem is to draw the best flowchart you can devise. It might improve your work habits, and spill over to improve your productivity and skill in your field.

PROBLEMS

1. Convert the following flowchart Arrow Method to a Left-to-Right Method (using connectors). See Figure 3.6.

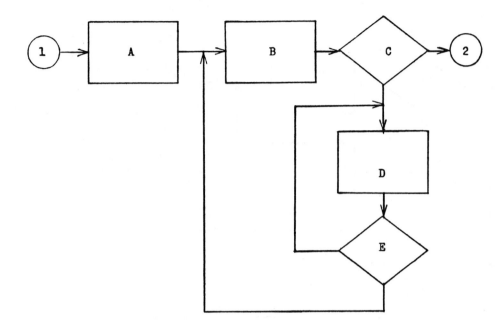

ANSWER

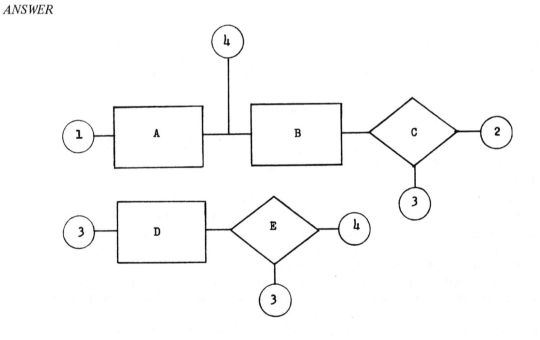

2. Convert the Left-to-Right Method (using connectors) to the Arrow Method. See Figure 3.6.

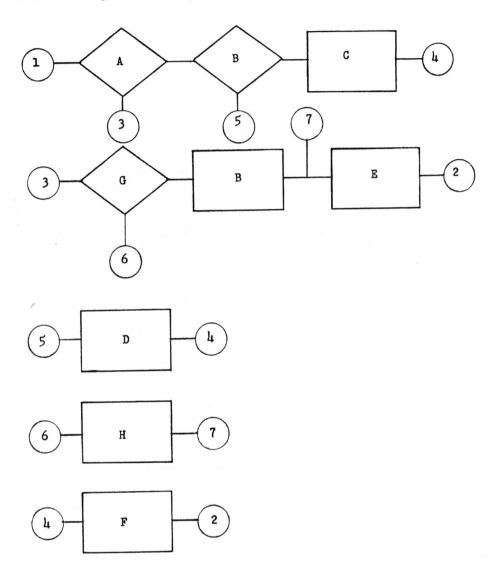

ANSWER

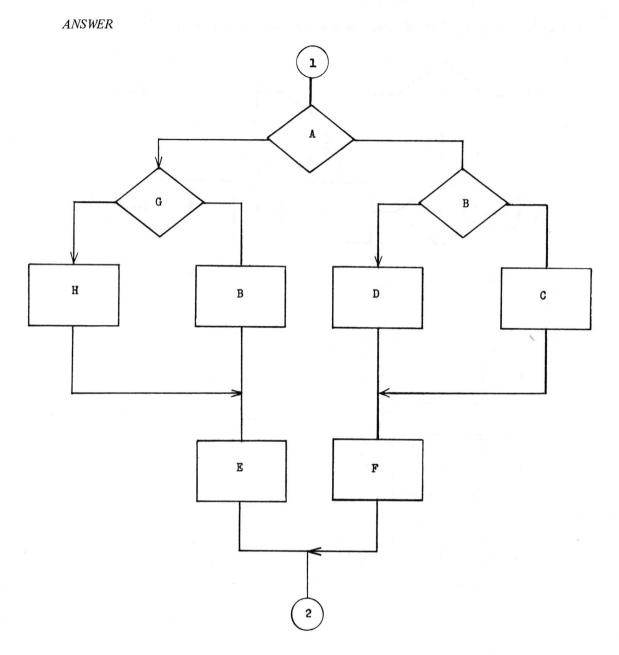

3. Draw a more efficient way of representing the flowchart below.

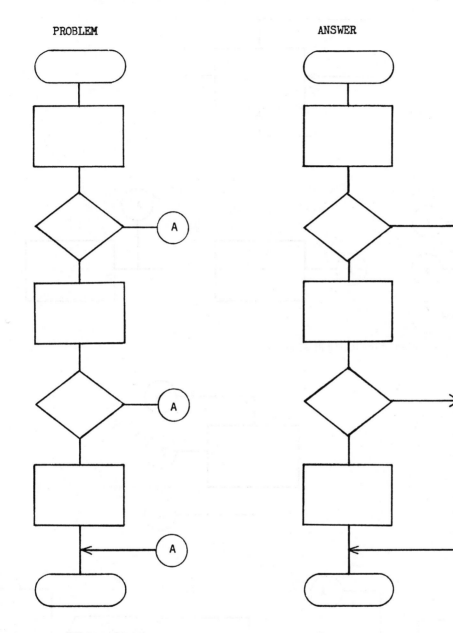

4. Correct the improper merging illustration.

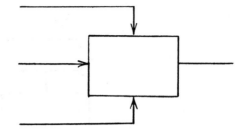

ANSWER

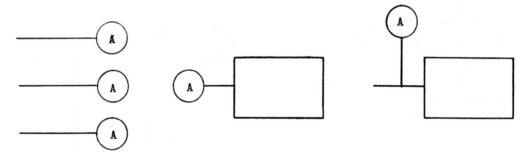

5. Correct the nonpreferred branching illustration.

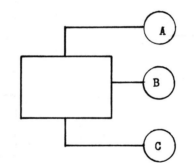

ANSWER

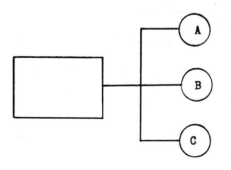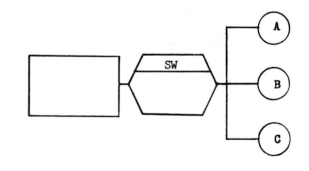

6. Connector symbols are used where space limitations require a flowchart to be sectionalized as shown.

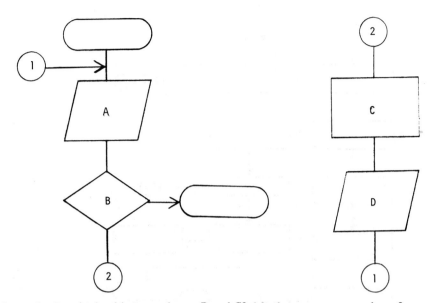

How is continuity obtained between boxes B and C? Are there any program loops?

ANSWER

Flow is maintained between boxes B and C using connector symbol (2) . *The program loops from box D to box A.*

7. Examine the flowchart* below. Comment on it in general and redraw it eliminating Connectors 2 and 3.

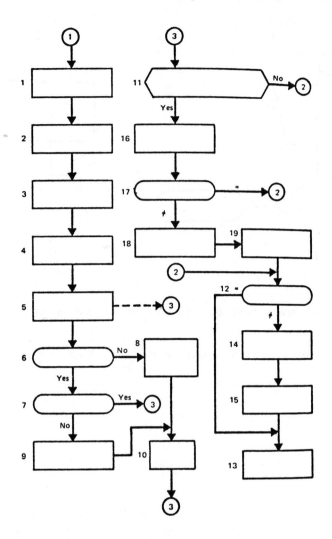

*Leo J. Cohen, *Operating System Analysis and Design,* (Hayden Book Company, Inc., 1970).

ANSWER

There are two different decision symbols used and neither one is the standard decision symbol. The main flow path visibility is not obvious. Box 5 has two outputs and the reader has difficulty in following the dashed line connecting the output of Box 5 to Box 11 regardless of the path taken by the solid line output of Box 5.

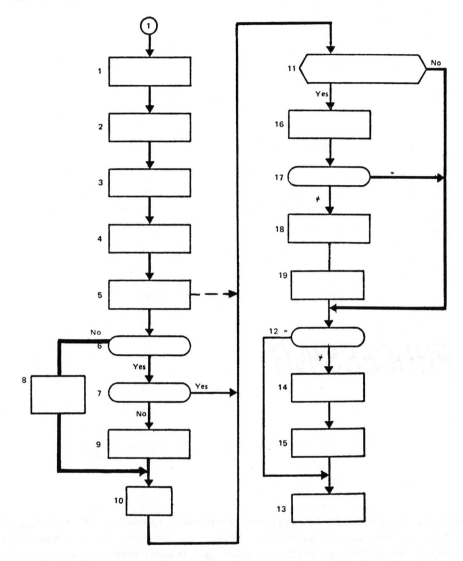

Additional flowchart symbols should be used to detail the significance of the dashed line output of Box 5 or the annotation symbol should be used to inform the reader of the associated details performed by Box 5.

4

DATA PROCESSING

The notion of sequential and simultaneous data processing operations is expanded here. Flowcharting permits a variety of interpretations for sequential, instantaneous, and concurrent operations. The time parameter is introduced in a flowchart of a program fragment to show execution time. It adds a third dimension to flowcharting. Before continuing, we examine the basic computer elements and configuration in terms of data flow for sequential and simultaneous operations.

COMPUTER CONFIGURATION

The computer is composed of the following major elements:

1. Input
2. Processor (Central Processing Unit)
3. Memory
4. Control
5. Output

The data processing operations are related to a basic computer (separate input/output) in the following manner.

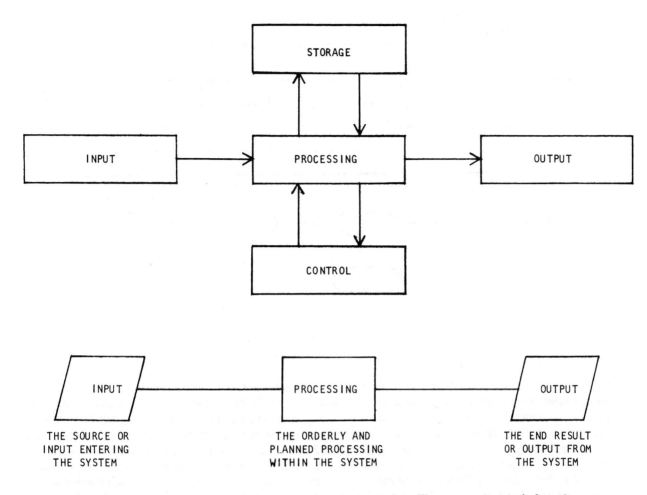

The CPU has the ability to perform certain operations on the input data. These operations include various arithmetic operations such as addition, subtraction, multiplication, division, rounding off and truncation of numbers, and taking square roots. Logical processing and decision making facilities are also included. Having the capability to process information, the computer must also be able to bring in new information to the processor when needed. This section of the computer, called the input, accommodates the input devices and their data formats and structures. In order for the computer to be useful, it must be able to furnish, in a form suitable for human use, the results of its processing to its output devices. The storage facility of a computer is not unlimited; it has some maximum capacity. Also, many computers have several types of storage, of different speeds and capacities (core, disc, drum, thin film, and solid state devices). The term "storage" refers to storage of input and output data, intermediate data results, and the means of communication between program segments, and the stored program itself. Control constitutes hardware control and software management of the operating system.

A very desirable feature often included in a computer is self-checking (i.e., the capability of the equipment to detect and indicate the occurrence of an error in its processing). Parity checking is applied to input/output processing functions and to core memory operations. Data processing checking schemes are software programs that check for: 1) nonvalid codes; 2) improper field configurations; and 3) transgressions of the allowed parameter range of data. Some checking schemes include error recovery programs.

SERIAL OPERATION

The serial management of computer components is quite limited in data processing capability. The machine must read (input) a single piece of data, process it, and write (output) it, before new data can be fetched.

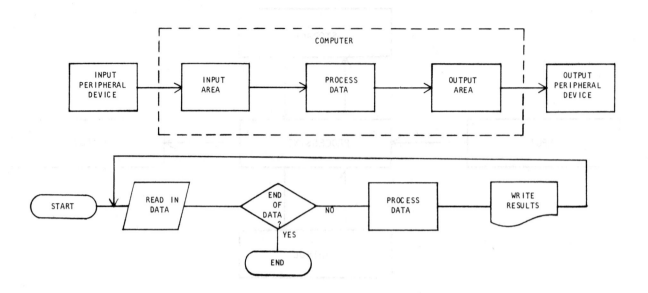

The diagram illustrates a serial operation in which only one operation is accomplished at a time. In some instances this may be desirable (e.g., where internal storage is limited). The storage would provide the capacity to store one data record and the program to process the information. The input and output area would be timeshared, since no new data would be read in until the current processing results were written out. The loop configuration is detailed in the next chapter. It is used here as a shorthand notation for representing the group chain, input, process, and output, as a repetitive data processing operation. Consider for the moment, a series of numbers are to be added. Normally, the numbers are lined up in column form according to some weighted position for each number symbol (e.g., units, tens, hundreds, etc.). The addition operation would proceed from top to bottom starting with the least significant column digit and continue from the top of the next left column.

The process would be repeated from right to left in a snake fashion until all columns were processed. If the same process were to be repeated automatically in a primitive fashion, the line coding would be a series of add operations from top to bottom for programming and the cards would be punched in serial fashion and computer entered in like manner. The first number would be added to the contents of the accumulation (after it was initialized or cleared to zero). The output would occur and for purposes here, it would be the subtotal. The process would be repeated for the next number to be added and outputted. The process would continue until the END card was entered and processed. All of the above are sequential operations that are ordered in a specific manner to obtain the desired result. Writing and repeating the same sequence of instructions for each addition and flowcharting the same is lengthy, time consuming, and confusing. The loop operation reduces the flowchart to a few flowchart symbols.

A loop operation with control ensures that a sequential operation is started, executed, completed and terminated as commanded in the example for an addition. The shorthand notation for addition can be represented very simply as indicated in the diagram.

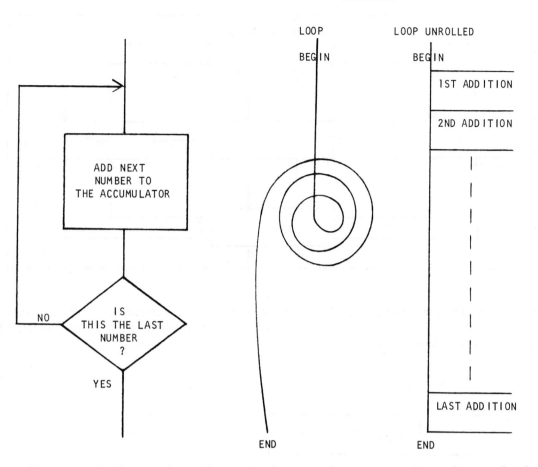

The graphic presentation shows a sequential operation that proceeds one step at a time under control without any possibility of error, neglecting abnormalities. The unrolling of the loop maintains the presentation of a series of segments in a straight line in which each segment (of equal length for this example) is the trace of one cycle of loop operation. The line is traversed in one direction (down), and no vacancies or discontinuities are permitted. The vacancies are removed by linking the terminus of one operational segment to the beginning of the next sequential operation. Every operation proceeds in descending order; reversing to repeat or process a previous numbered segment operation is not permitted. Sequential operations are no more than a series of operations uninterrupted from beginning to end or a series of line coding between two label identifiers for this example. A more sophisticated computer program would show parallelism and simultaneous operations such as those presented in the next section.

SIMULTANEOUS OPERATION

If the previous basic computer configuration is modified to permit each operation (input, process, output), to be executed independently, the diagram shown results.

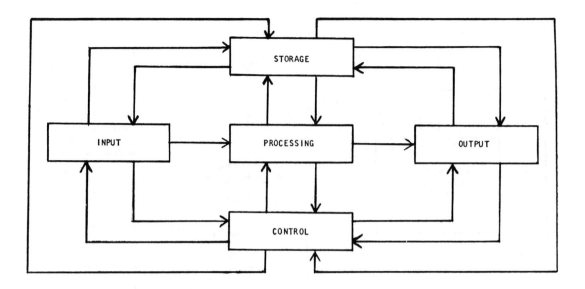

Again, the serial operational chain is retained. The data to be processed must be entered into the computer. Once this input data has been processed, the results are available for output. The serial chain of processing data (D) is presented as a time sequence described.

SERIAL OPERATION

SIMULTANEOUS OPERATION

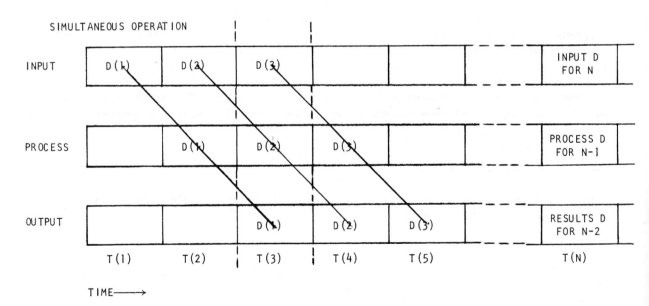

TIME⟶

Input, process, and output are shown as three separate and independent operations. The coordination among the three (synchronization) is maintained by the control subsystem and data correlation is accomplished by the software. The above diagram illustrates a simultaneous operation. Assume for the moment that each serial operation is executed in approximately the same time. Data, D(1), enters the system at time T(1). In the next unit of time D(1) is being processed while the next data, D(2), enters the system. Once the initial data, D(1), has been processed, it is now available at time T(3). Not until the third time frame can data, D(1), be outputted. Since each operation is independent and isolated from the other, each operation is executed independently of the other. In a typical time frame, N, the computer is performing three operations simultaneously. The storage facility in conjunction with the control subsystem maintains synchronization and the data processing is one continuous flow.

TIME

The three separate operations can be correlated and analyzed by flowcharting using the time parameter.

The flowchart sheets can be scaled, or the symbols can be scaled in terms of processing time, or time marks can be attached to the flowchart symbols in terms of processing units (e.g., machine cycle, memory cycle, gate transfer, etc.).

The time parameter introduces a third dimension to flowcharting. The reader should not confuse this three dimensional chart with that mentioned in an earlier description. The previous three dimensional flowchart could have implied the same time units as shown here, but the third dimension attribute contained the same operation with increasingly detailed description. It was called multi level of the same operation. The third dimension here correlates two or more separate and distinct operations being executed during the same interval of time.

Time granularity at an infinitesimal unit length is not implied. Flowcharts do not treat time granularity at the level of one pulse unit of the master timing system, resolution of gate priming and transfers, delays associated with device intrinsic storage time, and possible race conditions. The flowchart is not intended to replace the problem solving normally accomplished by the use of timing diagrams. In the usual flowchart the time granularity would be more at the macroinstruction level of detail and possibly exclude microinstruction granularity. The time unit would be the equivalent of total time or an integral unit of time (T) for an operation instead of being detailed as a sum of differential time units ($\triangle t$ or t) in a mathematical sense.

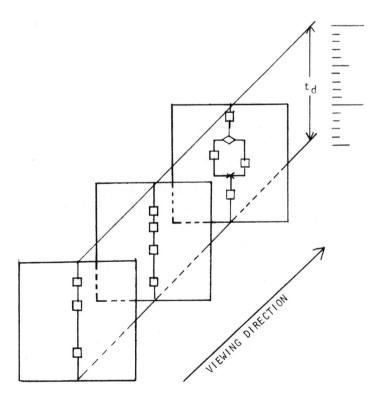

TIME ANALYSIS USING FLOWCHART

In most cases universal time units or relative time units are adequate to evaluate the fastest or slowest operation to modify a software design and obtain a more optimum solution. At other times, the time parameter indicates problem areas, and the technique is used for problem solution. If the flowcharts are drawn on transparencies and overlayed, the programmer can optically process the flowcharts and can evaluate his software design on a time basis.

SIMULATION USING THE TIME PARAMETER IN FLOWCHARTING

Once a model is defined and is to be simulated, the flowchart supplies visibility to the programmer to monitor the process. The time parameter is evoked and the third dimension is the projection of time upon each flowchart.

The flowchart is scaled so that time can be measured and correlated with each level of detail of the system being flowcharted. This is shown in the diagram.

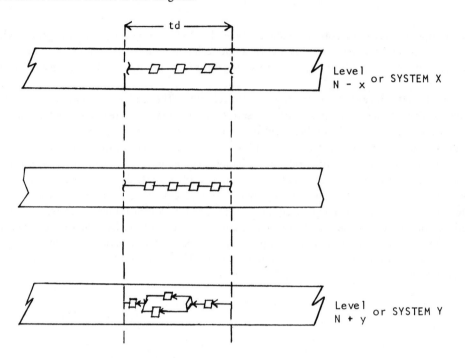

3-D FLOWCHART-TIMING ANALYSIS

In a simulation program, an event stack is used as a pseudo executive. The EVENT QUEUE order is in sequential time, and each event is processed as a series of operations in the following manner. Each event of the stack details those operations to be executed for this current event duration (e.g., analogous to a Digital Differential Analyzer (DDA) where one time frame is one revolution of a magnetic drum). After all the commands have been executed (other queues have been accessed, processed, and updated), the next time event is taken from EVENT QUEUE. The simulation clock keeps track of time as spelled out by the data content of the event word. The clock is used to operate the EVENT QUEUE. A time slice of concurrent operations for one time event is shown as a series of computations.

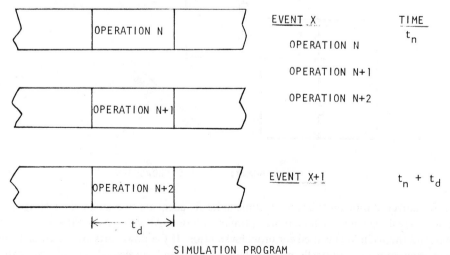

SIMULATION PROGRAM

If computation interdependence or data dependence is present, side effects will occur and degrade the simulation. Note the simulation presented here is a linear programming method.

Using the same model, the simulated operations are tracked in time using flowcharts. Each operation is flowcharted and time scaled as before on a sketch pad sheet (semitransparent). The sheets are then overlayed on a light table so that all operations for a given time event and duration can be viewed simultaneously. Any operation which does not meet its time budget can be discovered in this process of reading several flowchart sheets together (i.e., x-ray analysis). Once the problem is detected, the solution is obvious.

The timing analysis flowchart is used in analyzing the micrologic of a microinstruction of a computer operation in the hardware design phase. A timing analysis of a bit (or field) processing of a microprogramming instruction is another application of this technique. The microinstruction word field is drawn and serves as the start point of each logical process. The logical operations are drawn in column form (time axis), and each row depicts the simultaneity of the multiplicity of operations (see Figure 4.1).

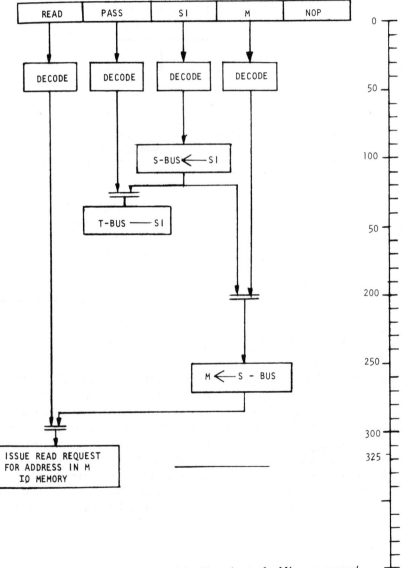

FIGURE 4.1   Flowchart of a Microcommand

PARALLEL MODE

The parallel mode symbol is used to represent two or more simultaneous operations. It can be used to represent the beginning or end of these operations. A typical simultaneous input or output operation is illustrated. A minicomputer has a limited or simplified set of controls to interface with peripheral devices. Once the transfer across the interface has been made, it is acknowledged after the parity has been checked to be correct. The minicomputer acknowledges the transfer, and the flowchart includes the operation of the peripheral device. The system does not include the verification of the Acknowledge for either transfer direction. For illustrative purposes, the minicomputer Acknowledge interface processing is flowcharted.

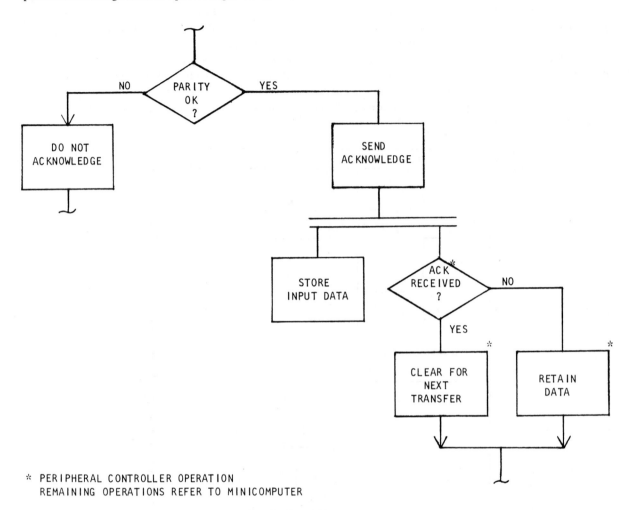

* PERIPHERAL CONTROLLER OPERATION
  REMAINING OPERATIONS REFER TO MINICOMPUTER

The flowchart illustrates a simultaneous operation of two subsystems: the minicomputer and its peripheral devices. The minicomputer acknowledges a valid transfer and the possibility of the peripheral device not receiving this acknowledgement is detailed. Since the peripheral controller is under direct control of the minicomputer, the data is not lost. The data is retained and is transferred on the next interrogation by the minicomputer. In this manner, whenever an Acknowledge is never sent, or is sent but not received, the peripheral controller retains the data until an Acknowledge is received under proper conditions and then clears its (peripheral) storage.

The parallel mode symbol for affording visibility to several operations occurring simultaneously is very effective. However, simultaneous presentation for a decoding process using the parallel mode symbol is not quite so effective and can even be misleading. Bit and field processing present problems when the parallel mode symbol is employed.

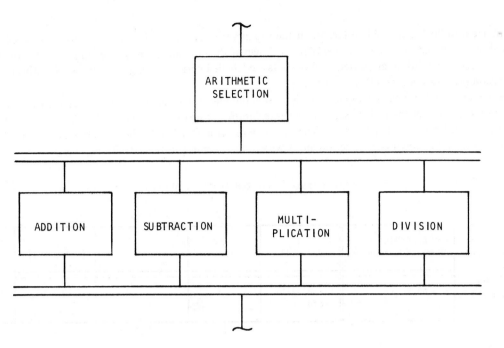

The arithmetic selection is a simultaneous operation where one of four possible codes is selected. The time required to select any one of the four is the same. Switch settings or code selection present a problem. In this example, decision making capabilities are required, and this is a misapplication of the parallel mode symbol. A preferable method of simultaneous operation of a matrix decode is a branch table.

BRANCH TABLE

The branch table is the equivalent of a two-column or two-row file structure. One row (or column) is used as a label identifier. The other row (or column) associates the location for the label contents. The selection of any row is a decision operation and with the selection, data flow continuity is maintained by processing automatically the contents of the location information. The previous arithmetic selection coding is shown in branch table form.

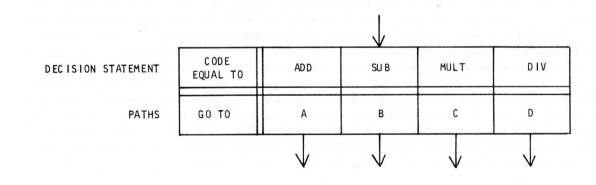

Entrance into the branch table is located at the top in a singular path. All four codes are simultaneously evaluated and the continuation of their logical paths is shown at the bottom. Each is independent of the other. Here all codes or decision conditions are stated, and outputs are labelled correspondingly. There is no confusion concerning the decision processes nor data flow continuity.

The branch table form can be applied to microprogramming. A microcommand or microinstruction exercises elementary control over the various system elements within a basic machine cycle. The machine cycle is now a universal unit of time for data processing. The time parameter is attached to the branch table symbol showing increasing time in the downward direction along with a microcommand field structure.

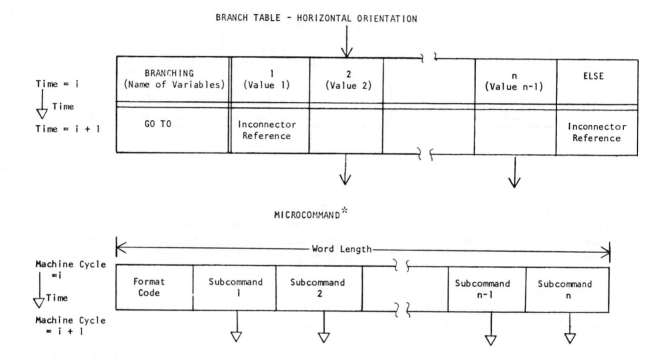

There are a number of subcommands included in one microcommand (or instruction). The format code or function code defines field location and lengths. It could include variable field formatting. The execution of the microcommand is the execution of each of the microinstructions sequenced by the control unit for each of the fields contained within a single microcommand (microprogram). All subcommand operations begin simultaneously. There are N parallel paths initially. Each field is executed under control of the timing subsystem but may not necessarily proceed at the same speed nor complete its operation at the same time. The number of parallel paths downward may be reduced by a merging operation. The opposite could also be true, so that at the completion of a given machine cycle, the number of total parallel paths may exceed the initial quantity. Also some operations are to be completed by a chain to the next microcommand, depending on the function field (format code). In terms of a machine cycle, the microcommand, initiation and completion, is executed over a finite time interval. A flowchart presentation of a microcommand generation is given in Figure 4.1.

DECISION TABLE

Normally, the chain list of decision symbols is sequentially processed or executed. However, there may be many applications where the entire decision chain is traversed in one time unit. A flowchart implementation of a hardware operation is illustrated here.

*A microcommand is an operation code for one or more micro-operations.

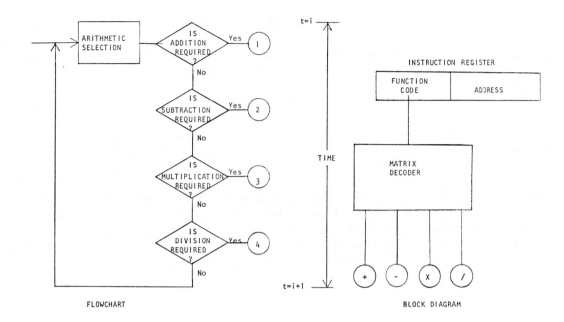

FLOWCHART

BLOCK DIAGRAM

Here, the arithmetic operation, when coded, is selected automatically. The decoding process is simultaneous for all codes and the selection of any of the four codes requires the same processing time.

SWITCH

The switching operation is quite similar to the previous examples. The hardware example will be repeated for a switching operation and the parallel mode symbol is included.

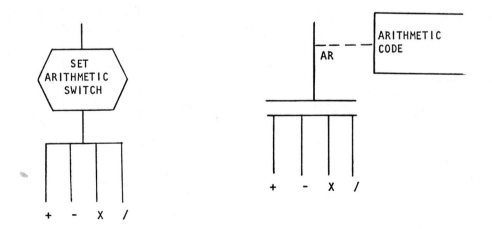

Concurrency (parallelism) as depicted here is not necessarily simultaneous. Instead, for a given time period, all the items are processed or executed implying simultaneity, and for all practicable purposes, that is what the flowchart conveys.

SUMMARY

The parallel mode symbol is used to depict simultaneous operations at a top level such as hardware and software system aspects. Gross detail visibility is obtained employing this symbol. In most cases the system concepts are adequate to formulate a solution, and the parallel mode symbol assists in conveying a total system configuration. Where decision making capability and other processing details are required, other flowchart symbols are preferred and recommended.

The branch table symbol, decision symbol, and switching operation can be used both for serial and simultaneous operations at the detailed flowchart level. The time parameter assigns execution time for a particular process. Time in conjunction with the three flowchart symbols can be used to express simultaneous operations that could be or would be interpreted as serial operations. If there is any possibility of misinterpretation, the annotation symbol may be helpful. Simultaneity can be obtained on a single sheet as two or more columns of flowchart symbols for detailed processing of an operation (e.g., ADD). This would be at the detailed elements themselves. Where separate and related operations are to be considered, each processing operation (or subsystem) can be flowcharted on separate sheets using the same time scale or equivalent units for each flowchart sheet. Time is used to relate or correlate several processing operations. For convenience, if the flowcharts are on transparencies and overlayed, several different flowchart operations can be evaluated in the time domain. Sequential data processing is a straight line approach. Simultaneous data processing is multitask, or parallel processing that would be more appropriately expressed in array notation.

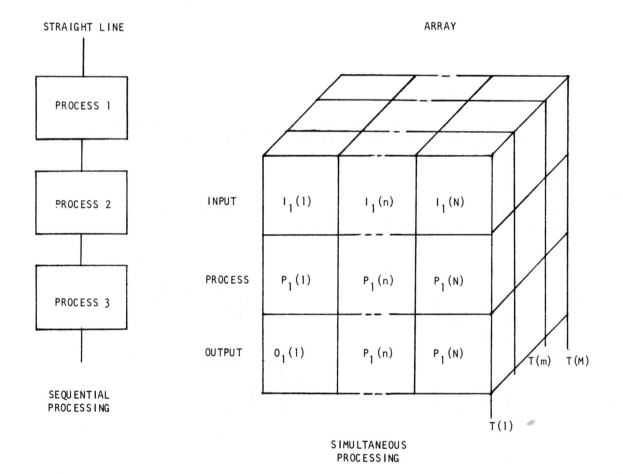

PROBLEMS

1. A programmer finds it difficult to express a return loop path using the standard flowchart symbol configuration shown.

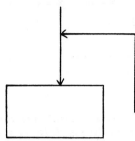

Instead he drew the following configuration to express his thoughts.

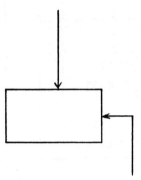

Examine both configurations and draw a flowchart symbol configuration that conforms to the standard flowchart practice and retains the programmer's information.

HINT: Consider the input to the processing block to be a table having multiple entries.

ANSWER

The programmer wants to express at least two separate and distinct inputs to the rectangle symbol. If it is sequential or parallel operation, the following method conveys this information.

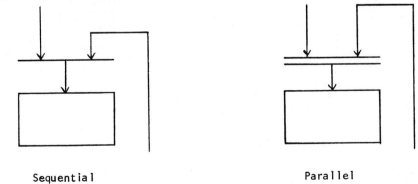

Sequential　　　　　　Parallel

Here two distinct inputs are expressed, and they are channeled (or bussed) into a common unit (table). Another parallel presentation could have the following configuration.

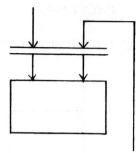

In the flowchart, the information is still retained. If in this instance the rectangle comprises many operations expressed in the encircled text material, the two inputs may have the following meaning.

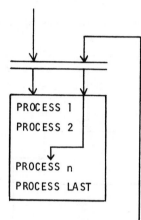

5

PROGRAMMING CONCEPTS

In the near future, algorithms written in flowchart language will be communicated directly into machine language (object code). A typical flowchart program schema requires five primitives as defined in Figure 5.1.

In examining these primitives, one discovers that the only primitive with more than one output is the branch/ decision type, and even here the two outputs for the decision type are related in the sense of defining the complete output domain: YES and NO (i.e., Yes, Not Yes), P and $\overline{P}$, and so on.

The symbol configurations for three basic programming operations (see Figure 5.2) that are language independent control structures retain the single input and single output concept expressed in modularization, and they form the basis for all structural programming.

Since flowcharts are natural models of representation, it follows that any program can be constructed from these three basic program structures, (using the five primitives). The reverse is also true (i.e., large programs can be reduced to represent these basic structures). A program of compound statements can be formed from simple statements, and conversely, compound statements can be reduced to simple statements.

The assembly of simple basic structured elements into complex programs (bottom up) and the reduction (decomposition) of complex programs to basic structured elements (top down) are shown in Figure 5.3. Fundamentally it is possible to duplicate the action of any flowchartable program by a program that uses the three basic program structures: 1) sequence; 2) selection (if condition); and 3) repeat (DO type).

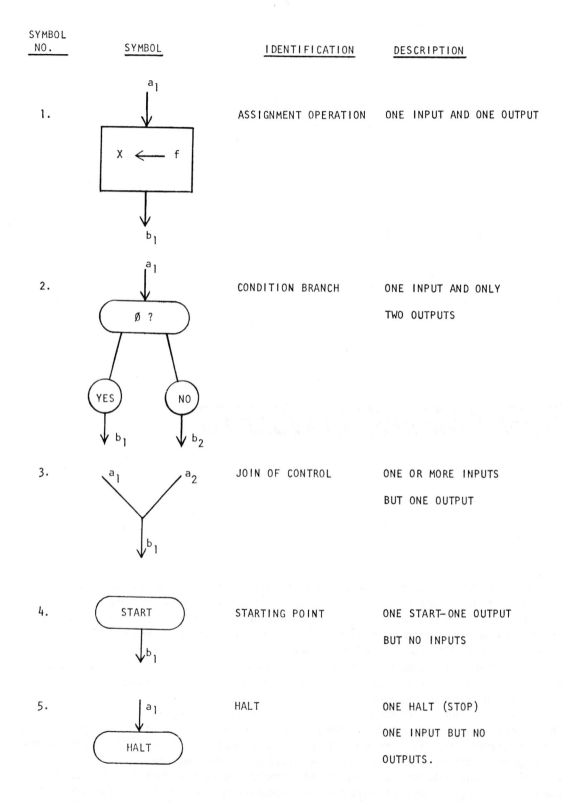

SYMBOL NO.	SYMBOL	IDENTIFICATION	DESCRIPTION
1.	a_1 / X ← f / b_1	ASSIGNMENT OPERATION	ONE INPUT AND ONE OUTPUT
2.	a_1 / Ø ? / YES NO / b_1 b_2	CONDITION BRANCH	ONE INPUT AND ONLY TWO OUTPUTS
3.	a_1 a_2 / b_1	JOIN OF CONTROL	ONE OR MORE INPUTS BUT ONE OUTPUT
4.	START / b_1	STARTING POINT	ONE START—ONE OUTPUT BUT NO INPUTS
5.	a_1 / HALT	HALT	ONE HALT (STOP) ONE INPUT BUT NO OUTPUTS.

FIGURE 5.1 *Five Primitives*[1]

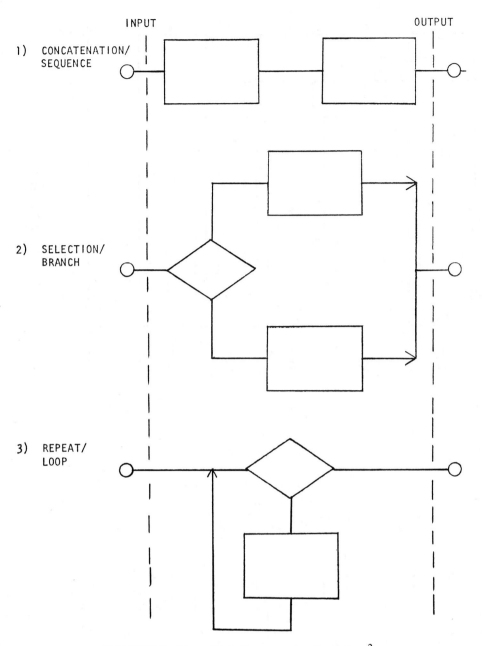

FIGURE 5.2   *Three Basic Programming Operations*[2]

The basics, five primitives and three structured elements, are expanded later to show the wide range of applications (e.g., program correctness, program equivalence, and theory of program schemas) that are dependent on these basics.

START

Since flowlines always connect one symbol to another, and some can terminate in mid-air (outconnector), some method of labeling is necessary to designate the starting point and the finish of a program. A terminal symbol enclosing the word "START" is therefore used to mark the beginning of the flowchart, and a similar one enclosing the word "END" is used to mark the end.

Normally there is no explicit start statement in a written program. One simply assumes that the first line is the beginning. This would imply that the job control language or the program execution is just taken for granted by a

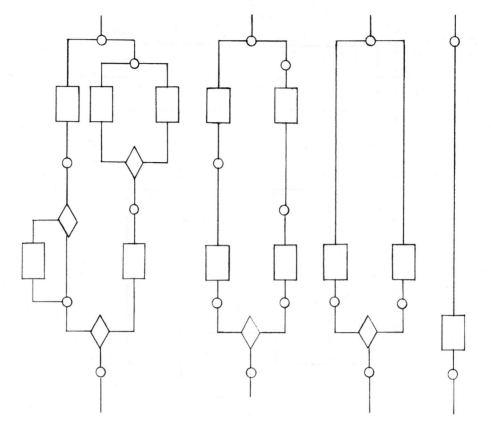

BOTTOM UP

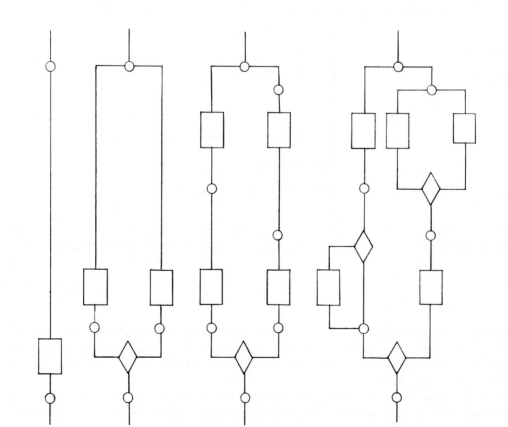

TOP DOWN

FIGURE 5.3 Flowchart Expansion and Decomposition

○— Segment End Point

series of control statements. This does not include the program itself. An examination of a typical program reveals certain initialization steps to be necessary (e.g., base register assignments). As a matter of fact, a series of assignment steps is necessary to establish the machine state prior to its utilization for problem solving.

INITIALIZATION – ENTRY

Initialization is the establishment of starting values to be operated on. Actually, initialization can be a repetitive process for establishing a set of conditions that includes component selection, device operation, and specific parameter values at the beginning of a processing operation. This procedure is necessary at the flowchart beginning and at any prescribed point prior to a loop operation, a decision, a switch, or a branch mechanization (see Figure 5.4). (Many subroutine link operations may require initialization prior to execution.)

Shown here are some typical examples of program starting operations which are also applicable to starting in flowchart drawing.

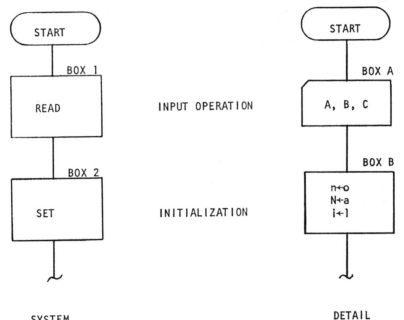

Boxes 1 and 2 are grouped together to show the availability of all values to be used and also the assignment of values for specific parameters and variables. In this instance, we are assured that the values and items are stored in advance by READing in the data, and initialization is the SET conditions for initial values. Just after the START terminal symbol it is common practice to consolidate and introduce all the input data as a precursor for program oepration. The reader is cautioned about the overworked and oversimplified state: READ. For our purposes here, it is adequate. In reality the addressing, operating, and formatting of data from input devices such as card readers, paper tape readers, and magnetic tape stations are all tasks not to be taken lightly, expecially when the tasks entail unpacking and assembling the data for computer entry.

The operation SET in Box B depicts three common choices for the initial values of parameters involved in an initialization process.

1. n in this example is given the value of zero (0).
2. N is made the value of a fixed number (a).
3. i is made the value of one (1).

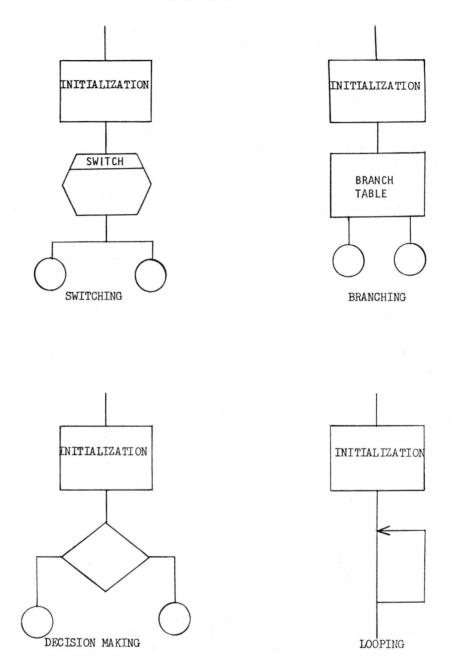

FIGURE 5.4   Initialization for Flowchart Operations

STOP

There are specific stop operations (END), but there can be numerous interim stop operations. A STOP statement terminates a program run. It may appear anywhere in a program. The END statement, which also terminates a program run, would appear as the last program statement (highest numbered line of the program) marking the very end of the program.

As stated earlier, a typical flowchart may contain several branching points. If all such branch paths do not converge on the same termination point (or merge with the same termination point), there will be more than one stop operation present in the flowchart.

CONTINUATION

More often, the information within the termination symbol may contain such words as: INTERRUPT; HALT; ERROR; DELAY; TEST; STOP; and EXIT. In essence, all result in the same action or inaction; that is the computer stops. Obviously, it will often be necessary, or desirable, for the operator to break in at this point to perform some function or action. But how? And furthermore, how do you return to the flowchart and where? Answers to these questions are not simple. In some cases, the return to computer operation is a lengthy process. The conditional state of the computer and the possibility of errors may not allow simple manual start procedures such as the pressing of an override or start pushbutton. For each type of interruption, a unique set of procedures is generally necessary to return the computer to an operational state. The annotation symbol in a flowchart affords a means for defining these steps and establishing the correct conditions. Re-entry may be uniquely defined by using the terminal symbol with the written word RETURN. Again, if there is more than one RETURN symbol in a chart, clarification is needed to prevent misinterpretation.

A common flowcharting error in start, stop, and continuation is the insertion of a print operation. Thus when the computer is interrupted or otherwise programmed to pause for operator cummunication, a printout is called for, and the computer continues its processing. The flowchart may have a termination point using STOP or an operation symbol designation, but no means of continuity is expressed for this stop-and-then-start operation. In some cases, the reader feels he is left in mid-air when he sees a print information within an Input/Output Symbol, or within the Manual Operation Symbol that is the final operation. In any event, the reader should be considered when terminating so abruptly. Because, the print operation is subject to misinterpretation as a possible end of the program, the terminal symbol is expressly defined for this purpose (see Figure 5.5).

PAUSE AND STOP FORTRAN STATEMENTS

The single word PAUSE statement is used whenever the execution of a program is to be halted temporarily (e.g., to permit the operator to set a switch). Execution is resumed by pressing the START key on the computer console. In contrast, the STOP statement brings the program to a permanent halt and terminates execution in such a way that depressing the START key has no effect.

Although this difference between PAUSE and STOP is widely established and accepted, the writer of flowcharts should not assume that his readers will be aware of the difference. He should format his flowchart to convey a continuation after a PAUSE, but not after an END.

PROCESS OPERATION — SEQUENCE STRUCTURE

Any processing, assignment or declaring statement can be charted using the process symbol. The process symbol is a simple rectangle, but its simplicity is lost if the rectangle is filled with complex text material. Fortunately the single complex rectangle can easily be replaced with a chain of rectangles with simple legends for greater overall simplicity. In the payroll example, a simple sequence was presented. Also the selection of the highest value of three numbers was presented as a series of operations. The series presentation (concatenation) is one of the basic program structures. Each process box can be subdivided (and can be constructed) using only the single entry single exit structures pictured.

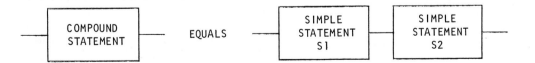

If statements S1 and S2 are simple statements, the concatenation of S1 and S2 is a statement (sequence). The process of generating a program flowchart representation of a system flowchart is based on sequence representation. A single process symbol on a system flowchart can actually represent an entire program and correspond to all of the symbols on the program flowchart for that program. This top down structuring is fundamental in multilevel flowcharts, and multilevel programming.

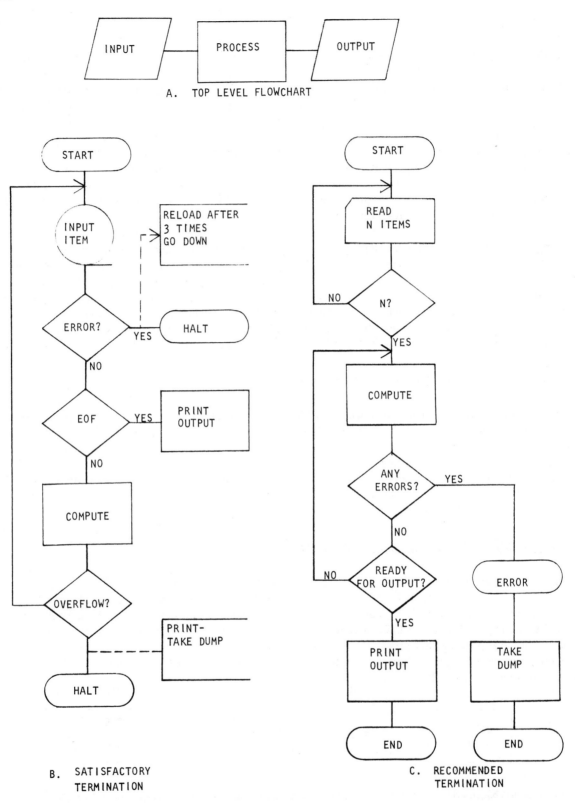

FIGURE 5.5 Stop-Print Error

BRANCH — SELECTION

The capability of modifying a sequence or selecting alternate paths through or around a sequence enhances computer utility and lends great versatility to control operations. Although the executed data control may not be known in advance, qualifying statements are made to control the operational paths when the input data is supplied. The methods and techniques of branching are numerous. The chart symbols that assist in portraying alternate paths (branching) are tabulated:

1. Decision Symbol
2. Connector Symbol
3. Branching Table
4. Switch Operation

Striping and connectors, used to detail operations elsewhere, are not a form of branching;

The decision symbol and the branching table enable the flowchart writer to identify alternate paths and include the conditions of transfer. The Connector and Switch Operation symbols may require auxiliary information in the form of processing operations (Process Symbol) and explanatory notes (Annotation Symbol) for clarification.

The decision symbol (defined in Chapter 2) expresses a certain relation between two current values. A basic statement that corresponds to a decision symbol is called an IF statement. Here is a typical example,

If A = B THEN C (where C is a Statement Number)

An IF statement has three parts, the word IF, to identify the statement; a logical relation which is either true or false (such as "A = B"); and the word THEN followed by a statement or line number (such as "C") to be executed when the relation has a logical value of TRUE. If not TRUE, it is automatically implied the next program set is executed. The relations that can be used in an IF statement are presented in Chapter 2. The purpose of the decision box is to cause a branch to another part of the program as a result of a test, or to control the choice between two (or more) alternate paths.

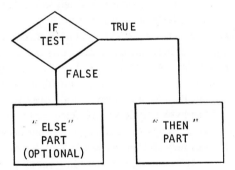

LOOP OPERATION

The ability of a computer to repeat the same operations with different data is a powerful tool which not only reduces the programming effort but can be represented in flowchart form succinctly. If, for the moment, we ignore the starting and stopping problems, the looping concept permits the union or fusion of the end path points to form a loop path for repetitive operation. The idea of a cyclic sequence of operations to be carried out repetitively is of fundamental importance, and is called an iteration loop. The iteration concept has a parallel in a high level language as a DO WHILE statement.

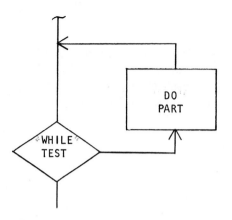

Once started, the loop, by definition, is endless unless there are means to exit from the loop. If one or more decisions are included in the loop and if they are not carefully planned, the computer program could recycle indefinitely. This occurrence is sometimes due to the program itself and at other times can occur for a given set of variable values and program structure peculiar to a computer installation.

The loop operation can comprise a series of loops, loops within loops, or a mixture of both. However, each individual loop, whether a part of a series chain or nested, must possess the following characteristics:

1. One or more operations (as a rule).
2. One or more decisions or testing procedures.
3. Incrementation, modification, or indexing.
4. Entry (initialization) at a qualified point.
5. Exit (see Item 2 above).

If a loop exhibits these five characteristics, it will perform correctly. Any ordered sequence of the five is acceptable. The order is irrelevant as long as each characteristic is permitted to function as intended and is not abrogated by the action of another operation within the loop or loop structure.*

When flowcharting an iteration loop, it is necessary that the chart be structured to show all the details necessary to validate the loop operation. In a good flowchart, the chief structural aspects of the process (e.g., the loop) will stand out. Some typical loop structures are shown in Figure 5.6. In following any of the loops symbolically, the validity of any iteration must be logically correct for insuring an exit from the loop.

Figure 5.7 shows a basic loop operation (Part A), an improper loop operation (Part B), and a more detailed loop operation of Part B (Part C). Figure 5.7B shows an endless loop. Once started, it will calculate the sum of A and B and output the data indefinitely as indicated. In reality, this may not occur because the data source (card reader, paper tape reader, or magnetic tape station) may contain sensors to indicate the existence and exhaustion of the media supply. A simple solution with effective control, the decision symbol, offers the means of either recycling or exiting from the loop. Indeed the decision symbol is essential to exiting from a loop operation. The loop addition example in Figure 5.8 illustrates the versatility and application of the decision symbol. In Figure 5.8A, a boundary condition is established as N. If the test of N is not met, the appropriate loop direction is taken and includes one or more calculations (Boxes A and B) that can change variable n to a value that permits loop exiting. In Part B of the same illustration, a test is made to see if the difference of two numbers is zero (0) or if equality exists. Part C is an indexing method. The loop operation is repeated from one to N. A tabulation printout will result in an ascending form for all the values of C.

At this point, it is appropriate to examine the structure, looping, exiting, and conditions of initialization with a simple example. Recalling that each of the five loop characteristics must be present and not nullified by another operation, we show a typical initialization error in Figure 5.9. The error made here is that the initialization operation is included in the loop operation. Obviously, the initialization is essential in defining the conditions of loop entry, but not at the expense of nullifying the modification of n (back to zero). Accordingly, a little thought should quickly make it clear that the resulting program will never exit, because one, and only one, operation using identical values for each cycle will be repeated. In most cases loop construction errors are not as obvious as the

*Francis K. Walnut, *Introduction to Computer Programming and Coding* (Prentice-Hall, 1968), p. 158.

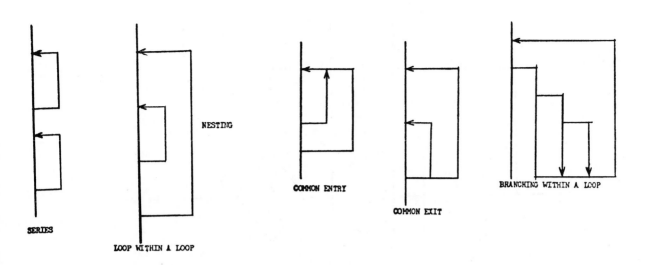

FIGURE 5.6   Various Loop Structures

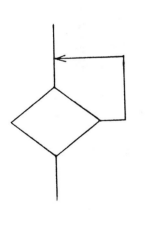

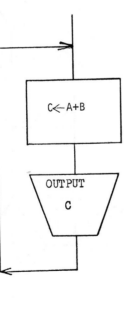

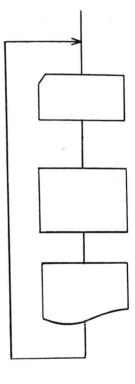

A   B   C

FIGURE 5.7   Improper Loop Configurations

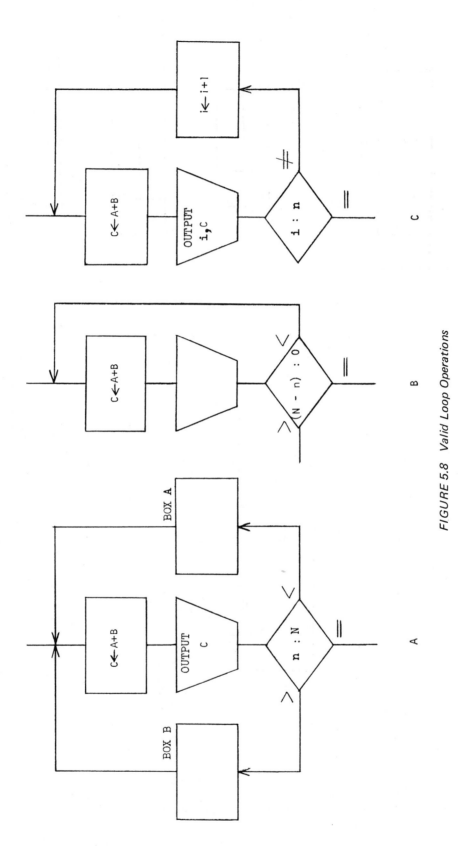

FIGURE 5.8   Valid Loop Operations

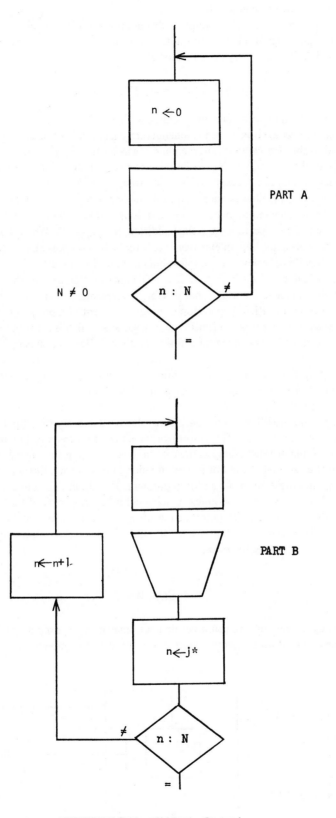

FIGURE 5.9   Invalid Loop Operations

one shown here. Many are due to programming and variable assignments. It is always preferable to flowchart the problem, examine it, and then implement it.

A loop example is presented here for a simple addition problem. (The variable subscript notation is used here while the high level language (FORTRAN, etc.) notation is introduced in the next chapter.) Let us say that n numbers are to be added. The sum may be expressed as:

$$S = \Sigma \, _{i=1}^{n} \, X_i = X_1 + X_2 + X_3 + \ldots + X_n$$

The flowchart for performing this addition is shown in Figure 5.10. Initially, the sum S is set to zero and subscript i is set to one, in Box 2. (Box 1 has been deliberately ignored for this description.) In Box 3, the current sum of zero is added to the first value X_i. Remember i was set to 1 before entering the loop. On leaving Box 3, after the first addition, the sum equals X_1. In Box 4, a test is made to see if the addition has been completed. For the moment, i equals 1 and not the value of n. Therefore, we go back to Box 3 via Box 5. Before the loop is closed, the current value of i (1) has a one added to it, making i equal 2 on leaving Box 5. Every time Box 3 is entered, i will have been increased by one. The cycle or loop is repeated until i + 1 equals n (i.e., the updated value of i equals n, and Box 3 performs the final addition $S + X_n = S$). When the test is performed in Box 4, an affirmative answer is obtained and an exit from the loop to Box 6 takes place. The sum is printed and the operation ends. It is worthwhile to review the complete operation in finer detail.

First, X is both a variable and a subscript variable. The initial assignment of the variable defines the variable, but the assignment must precede the use of that variable as an element in any operation. Defining the variable assignment is performed by the READ operation or input statement of a programming operation. Particular care must be exercised to ensure the proper definition of any variable which is subscripted with a second variable; example X_i. First, the variable itself must be defined and then i. Here the subscript was made greater than zero, but not greater than the value of n. The decision to exit from a loop is based on some means of control by the programmer. The appropriate technique is quite simple. The subscript i was introduced for the express purpose of counting the number of repetitive cycles to perform the addition.

The actual count started off by initializing i to one. Each previous total is retained and incremented by updating. The counting is monitored by incrementing its value by one every time the loop is cycled, and when i reaches the value n, the looping ends. This counting procedure keeps track of the number of loop cycles and can be reported out as indexing or referencing, as the case may be. By requiring the value of i to be part of the output, it must be treated as a variable and not just as an index subscript for control purposes in FORTRAN. This would require coding modification of the addition problem. This should not cause any confusion, because from a flowcharting point of view i serves as an indexing scheme for keeping track of the operation. An alternate flowchart is given in Figure 5.10, with a different location for indexing (counting) and a different initialization value for the index variable. This technique of counting can just as well be accomplished by decrementing the limit by one for each loop cycle and testing for zero (0).

SUMMARY

There is a commonality among the five primitive elements and the three program control structure elements. Each is based on a definable unit. Both concepts require a structured unit or module that has the following description that can be specified.

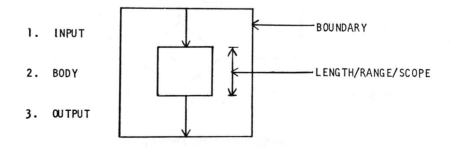

1. INPUT

2. BODY

3. OUTPUT

BOUNDARY

LENGTH/RANGE/SCOPE

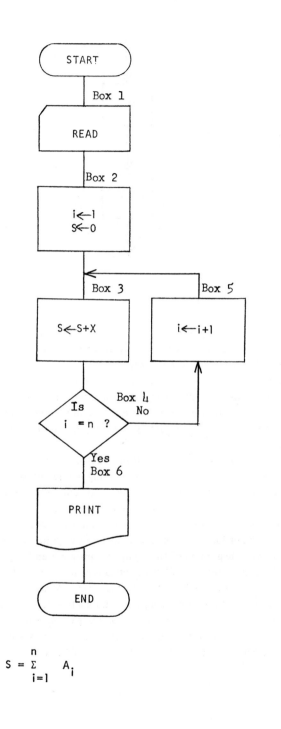

$$S = \sum_{i=1}^{n} A_i$$

Method Discussed In Text

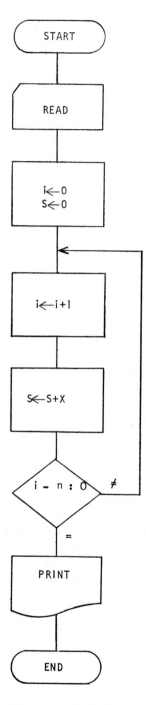

Alternate Method

FIGURE 5.10 Number Addition

It should be noted (in terms of the flowchart form) that each structure as a single entity contains one input arrow and one output arrow. The same is true for the rectangular boxes appearing within each structure.

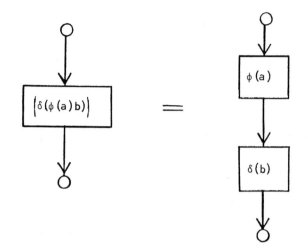

Thus it is possible to replace a rectangular symbol by any program operation structure or module from Figure 5.2 (viz., concatenation, selection, and repeat). By doing this recursively an arbitrary number of times, one can build up a structure of arbitrary complexity (see Figure 5.3 left side).

The primitive attributes of the basic programming structured elements enable module testing. Each of the three modules is constructed to have a single input and a single output. The output can be completely defined in terms of its input and its internal processing operation. Therefore, if the output can be completely defined or predicted on an element basis, the module can be defined and specified completely. Under this condition, the module can be tested in accordance with its specification.

The sequence structure permits chaining and multileveling. The selection structure permits tree configuration and program modification. The loop structure permits the elimination of chaining where appropriate by an indexing scheme or a conditional test where an adjusting process is included in the loop. The loop comprises five items in any order. The addition illustration shows the indexing in the forward and return path. The multiplication problem shows the test is performed upon entry into the loop and after the computation is made.

The START symbol represents the beginning operation which is most associated with input data, initialization, and establishing a specific environment. For the very beginning of the computer program and any subsequent introduction of a routine or loop operation, the equivalent START function is repeated, explicitly or implicitly.

For every START there is a corresponding STOP. The STOP may terminate an operation and is used to signify the beginning of the next or it is the final END; no continuation. Stopping is associated with the clean up details and outputting the results whether it be for immediate use (printer) or for archiving (magnetic tape). Between the START (beginning) and the STOP (end) some type of transformation, conversion, or process is performed to connect one with the other to form a primitive unit where center part (body) has one of three program control structure elements.

REFERENCES

1. Floyd, Robert W. "Assign Meaning to Programs" *Proceedings of Symposium in Applied Mathematics of the American Mathematical Society,* Vol. XIX, pp. 19-32, April 1966.

2. Yelowitz, Laurence, "Toward Optimal Paginations For Structured Programs," *Symposium on Computers and Automatia,* p. 227, Polytechnic Institute of Brooklyn, 1971.

3. Dahl, O. J.; Dijkstra, E. N.; Hoare, C. A. R. *Structured Programming,* p. 1-72, London and New York: Academic Press, 1972.

PROBLEMS

1. A programmed data transfer is performed by the I/O transfer instruction. It can be identified as unconditional transfer, conditional transfer, or program interrupt. Depict the above three operations in an elementary manner using flowchart symbols.

ANSWER

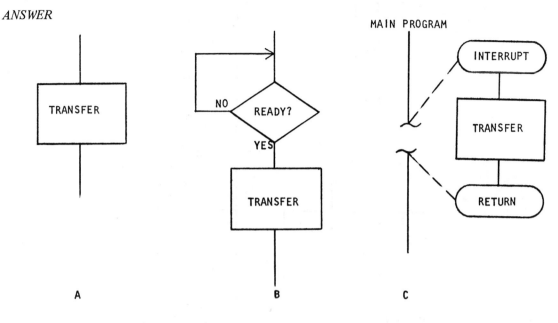

Programmed I/O transfer: A — unconditional, B — conditional, C — program interrupt.

2. Consider an entry point in a total program at the very beginning, at the very end, and any location between these terminal points.

    Hint: Use a table to identify the input, operation, and output for the beginning, middle, and end segments respectively.

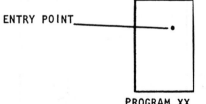

PROGRAM XX

ANSWER

	VERY BEGINNING	VERY END	MIDDLE
INPUT	Input Data is Zero		
OPERATION	START	STOP	
OUTPUT		Output Data is Zero	

3. Insert the decision symbol appropriately in the following figures to permit valid loop operations.

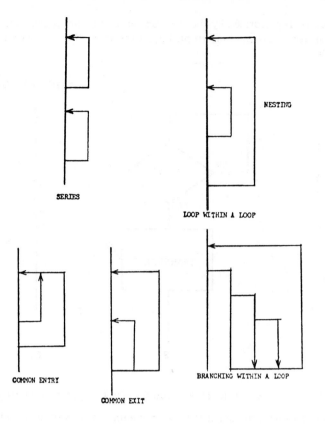

ANSWER

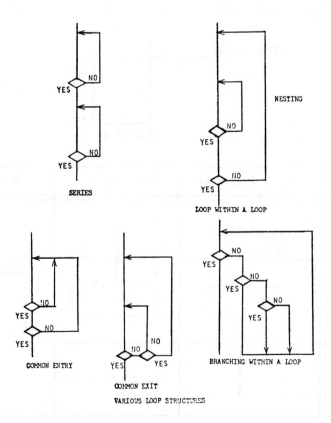

4. Connectors in the path of a loop should be avoided because they make the existence of the loop less obvious to the reader. Rearrange this flowchart to show the loop configuration.

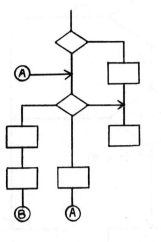

ANSWER

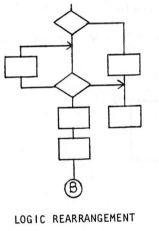

LOGIC REARRANGEMENT

5. The range of a variable is given as follows:

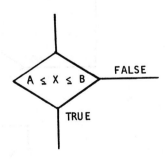

where A and B are Constants and X is a variable

Expand the complex test into simple conditional tests.

ANSWER

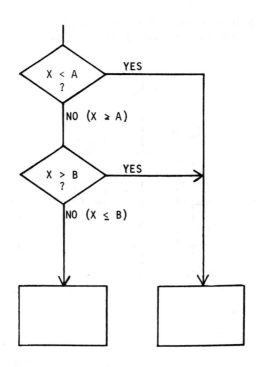

6. Express the series multiplication in loop form.

$$PROD = \prod_{i=1}^{n} = Y_i = Y_1 \cdot Y_2 \ldots Y_{n-1} \cdot Y_n$$

ANSWER (Two alternate solutions are shown.)

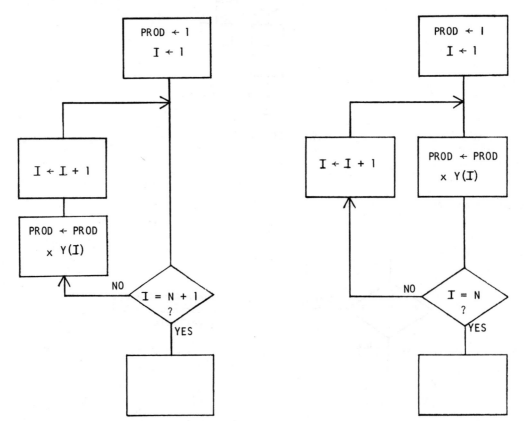

See Chapter 6 for explanation of notation used here.

7. Flowchart the selection of the largest of three numbers in a loop configuration. (In Chapter 3, the problem was without a loop.)

ANSWER

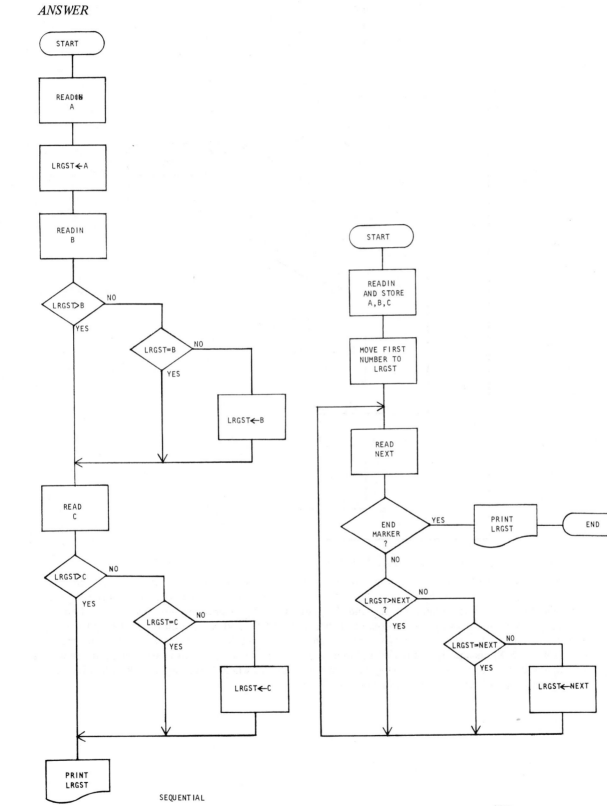

8. Flowchart the algorithm to find the largest number of the total input of N <u>positive</u> numbers defined as X_i (where $i = 1, 2, \ldots (N)$). Decrement the index $(i - 1)$ for each loop operation.

ANSWER

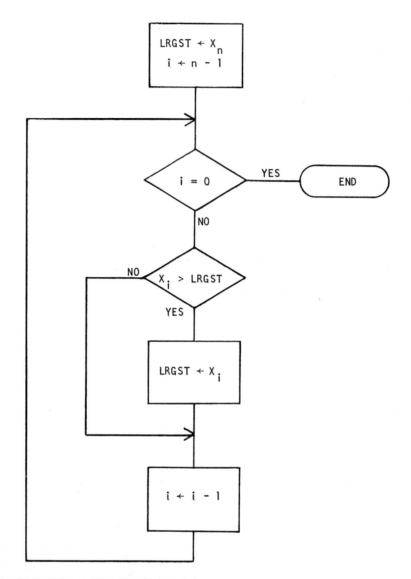

FLOWCHART DESCRIPTION FOR PROBLEM 8

The input (not shown) is presently stored in memory. The first box prior to loop entry initializes the index i ($i \leftarrow n - 1$) and location LRGST is set to X_n (LRGST $\leftarrow X_n$) and total number of inputs (n) is part of the input data. The first comparison is between X_n and X_{n-1} of where X_n is presently residing in LRGST to begin the process. Next in succession the comparison is between LRGST and X_{n-2} followed by LRGST and X_{n-3} and so forth. LRGST always contains the largest item value after each comparison. Index i begins from $(n - 1)$ because X_n is already in LRGST and there are only $(n - 1)$ comparisons to be made. The index i is decremented after each comparison. The loop process terminates when i reaches zero. With this index scheme, the test constant is always the same, namely zero.

6

PROGRAM FLOWCHARTING

The material covered in this chapter is an extension of the previous chapter. The challenge for the reader lies in learning to describe clearly a series of steps constituting a program segment. The emphasis is on putting together software segments and not on developing and solving equations. There is no intent to teach programming procedures nor programming languages, but the reader will discover that the information in this chapter can assist him in mastering those subjects.

Number, variable representation, and array notation are techniques of identifying and labelling that are common in programming. Looping, branching, and switching are basic operations and essential for a programming language to have utility. Hence, a flowchart of these operations may be documented independently of programming language or type of machine. The reader is introduced to high level programming languages via flowcharting building blocks that implement programming language statements from ALGOL, COBOL, FORTRAN IV and PL/I.

It is not intended to flowchart (i.e., document) an existing software program. However, there are cross referencing techniques between flowcharts and computer programs, a few of which are presented here. These techniques enable one to organize his thoughts by flowcharting the plan of attack on a problem, optimizing the solution, and as an auxiliary byproduct using the flowchart to check the implementation for performance accuracy. The introduction of high level languages presents a method of flowchart implementation. In some cases, the subject of programming languages is presented in flowchart form for comprehension.

VARIABLE NOTATION

It is important to understand the relation between the labels and the values of variables. A quantity may be referred to by its value, or by some representative designation such as a name or symbol. Thus, the number two (2) may simply be called TWO while its label naming is not quite so clear. On the other hand, if one knows that 1.4142 is the square root of two, he may prefer to refer to it as SQRT2 or SQRTWO. Anyone who has read a few programs has encountered some exotic names for variables. In most cases there are no formation rules except for the first permitted symbol and length of symbols as specified by the high level language. In FORTRAN the first character must be alphabetic (A through Z) and the maximum number of characters depends on the processor used.

The name or label of a variable (or constant, for that matter) and its quantity or value are two separate and distinct entities. The example of TWO for the value of 2 is a convenience, and so are INCM, TAX, LOG, and other strings of letters and numbers which assist the programmer in sight reading his program and correlating the logic flow with his program and flowchart. The label is used to refer to a physical location such as: memory address; hardware register; tape storage location; disc storage location; input on a punch card; or punched paper tape. The value of the name (variable) is the value stored at the location at any particular instant during the data processing. If the value is fixed, finite, and remains unaltered during the processing operation, it is considered a constant. Thus, the quantities 2 and π are constants. Moreover, if the location for storing the value of 2 or π is read from memory only and is never replaced by another value or if the location is not used for temporary storage, then obviously the locations represented by TWO and Pi are also constants.

Once it is agreed that a name (variable) can be used to represent a specific memory location and/or a means of addressing information, the variable notation can be extended to address a set, group, or class of data in a number of physical configurations. The convenience arises through the use of a single name to represent an entire list of values or an individual member of the list. The method employed is called subscripted variables. Note that the subscript may be either a constant (integer) or another variable. The subscripted variables constituting a list or table are stored in a contiguous manner. Mathematical artifices permit a number of configurations by an indexing scheme to associate, on a one-to-one basis, each value to a position or location as shown in Figure 6.1(A).

It is often convenient to treat many items of data simultaneously as a single entity. This is done by using an array notation. To reference an item or element of an array, one uses one or more subscripts to indicate its position in the array. These position numbers are called indices.

An array whose elements are selected by one index is called a vector. It has one coordinate and can be thought of as a collection of items arranged in a line (horizontally or vertically). An array whose elements are selected by two indices is called a matrix. It has two coordinates and can be thought of as a collection of items arranged in a rectangle (i.e., surface). An array whose elements are selected by N indices is called an N dimensional array. It has N coordinates and can be thought of as a collection of items arranged along N mutually orthogonal coordinate axes. Figure 6.1(A) shows a vector, a matrix, and a 3-D array, all with five elements in each dimension.

Although most arrays have more than one element, arrays of a single element are called scalar quantities. Since each name is unique and a single element is present, the name suffices as the identifier and no subscript is required. A single element has no coordinates and can be thought of as a point. The number of indices required to specify positions in an array is given by its rank. The dimensions and ranks of scalars, vectors, matrices, and arrays (rank-3) are shown in Figure 6.1(B) together with the programming and mathematical notation.

The mathematical notation for describing the members of a single subscripted variable is:

$$\overset{n}{\underset{i=1}{X_i}} = X_1, X_2, X_3, \ldots, X_n$$

The index is a means of associating a value to a name on a one-to-one basis. This notation is slightly modified for programming purposes. Instead of being written as subscripts, a specific value (index) must be enclosed in parentheses (square brackets [ ]) following the name:

$$X(1), \ X(2), \ X(3), \ \ldots, X(N) \quad \text{FORTRAN}$$

$$X[1], \ X[2], \ X[3], \ \ldots, X[N] \quad \text{ALGOL}$$

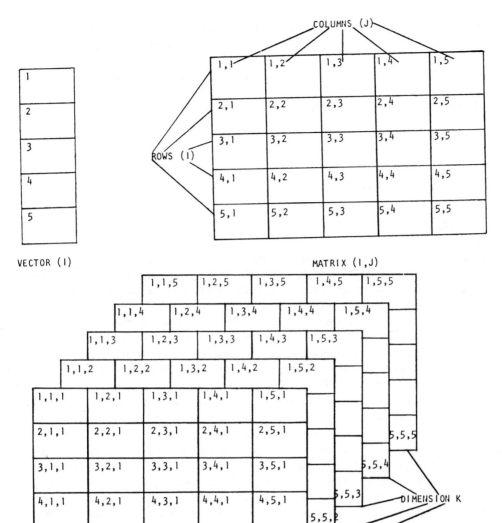

FIGURE 6.1(A)   Subscripted Variable

	Scalar	Vector	Matrix	3-Array
Geometric Interpretation	Point	Line	Surface	Volume
Number of Dimensions	None	One	Two	Three
Rank/Order	0	1	2	3
Programming	Name	Name (I)	Name (I, J)	Name (I, J, K)
Mathematical	X	X_i	$X_{i,j}$	$X_{i,j,k}$

FIGURE 6.1(B)   Subscripted Variable

Because the parentheses clearly delimit the extent of the subscript, any expression (provided that it represents an integer value) may be used or computed to select an index of the variable:

$$X\,(3)\quad X\,(I+1)\quad X\,(\text{Variable})\quad X\,(\text{Computed Value})$$

When multiple subscripts are present within the parentheses, commas are used as delimiters between each subscript.

$$X\,(I+1, J)\quad X\,(I, J, K)\quad X\,(I, A+B, C\times D+E)$$

Note: The value of A+B and C×D+E are integers.

COUNTING AND INDEXING

Before some software operations can be presented, some precursor information is necessary. In the previous section, the subscripted variable for counting was introduced, while indexing operations are discussed in this section. These two items, subscripted variable notation and indexing, are essential data processing tools and they can simplify the description of lengthy or complex operations.

The subscripted variable presents a powerful concept. Single subscripted variable processing is an invaluable aid in solving scientific problems. Now, the mathematical concept of surface and volume configurations permit dimensional solutions and operations in mundane areas: optical scanning, processing, and signal enhancement; three dimensional graphics and other video presentations; solution of ecology problems such as flows of water and air (weather forecasting); space management for inventory control; positioning of pipes and electrical conduits (three dimensional) in new building construction; and so on. The concept of more than three dimensional space is difficult to envision, but programming permits N dimensional system processing.* The commonality of multisubscripted variables is the sequential processing order and the nested loop configuration.

In cycling and loop iterations, the technique of stopping and exiting from the loop is based on some characteristic of the variable which also serves the purpose of the loop. The appropriate technique is quite simple: a new variable is introduced into the program for the specific purpose of counting the number of iterations performed. The actual count is maintained by simply adding one to the variable each time the programming goes around the loop. The decision to leave the loop is based on a preset final count (n) and the counting variable i. As a rule, the programmer assigns a value to n and the variable i can be incremented by one or more. The single subscripted variable (vector) is an implementation of this schema. The subscript magnitude terminates the loop and the subscript itself is used as an indexing operation. It may already have occurred to the reader that the increment (one or more) can be used for searching and retrieving information in file processing. A file configured as a fixed record total (the record length may be fixed or variable) can be scanned and data retrieved in terms of physical positions by incrementing. Obviously, files can be generated and ordered in a similar manner. A subscripted variable nested loop operation is shown in Figure 6.2.

LOOP CONTROL AND TRANSFER OPERATIONS

In problem solving, the software system designer has numerous algorithms and techniques with which to implement his ideas. Once a problem is defined and guidelines are established, the software system designer specifies his requirements in a similar manner; sketching out the system by drawing a flowchart. The flowchart outlines the data handling, logical, and computational requirements. It can also depict the software and hardware configurations. The flowchart is drawn as a method of solution and is likely to be under constant revision until a solution is acceptable or a consensus of opinion dictates a preferred direction. The programmer translates the software flowchart design into an available programming language. The programmer will flowchart the system requirement (program flowchart) and check it out before coding is initiated. Later, the program flowchart serves as part of the program's documentation, because it is usually easier to understand than the program listing (coding and comments) itself.

*In a few cases the programming language may limit the number of subscripts. One version of FORTRAN, for example, limits the number of subscripts to seven, but other versions set no limit.

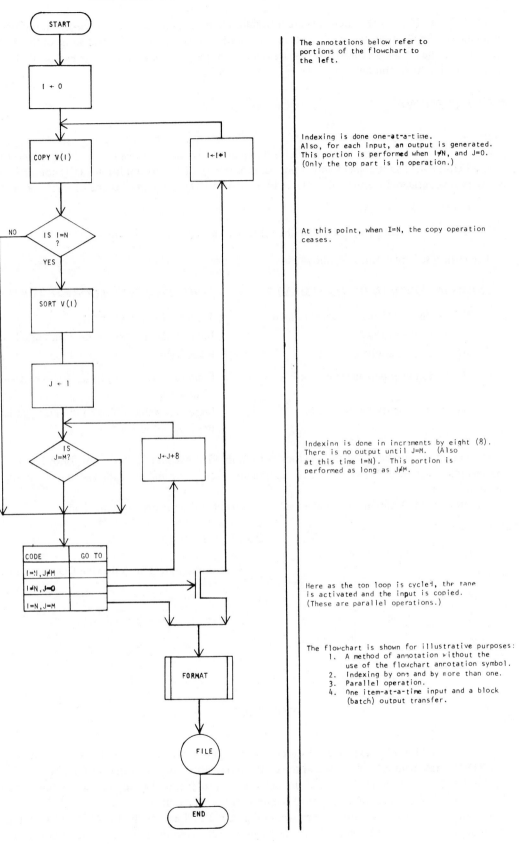

FIGURE 6.2 Scan and Sort

The DO, GO TO, and IF statements are introduced here to give the reader an insight into programming, high level language application, the meta-language connotation of these languages, and some of the methods of cross referencing. The reader is expected to research the literature of programming languages for details: structure; statements; function; and notation.[2, 3]

DO STATEMENT

Every procedural language contains a statement of the DO type.[1] Without question, the DO type statement is a powerful one, since it greatly facilitates definition and control of repetitive processes. Consider, for example, the simple FORTRAN DO statement with a normal program exit (i.e., no conditional (IF) nor unconditional (GO TO) transfers are considered for the moment). A FORTRAN[4] DO statement uses one of these forms:

1. DO N i = M_1, M_2, M_3

2. DO N i = M_1, M_2 (where M_3 has the value of one when not explicitly stated)

The terms used are defined as follows.

TERMS IN FORTRAN DO STATEMENT	CORRESPONDING LOOP PARAMETERS
1. N = number (label) of last statement	Terminal (end) statement.
2. i = control variable	Index variable whose value is modified per loop operation.
3. M_1 = initial parameter	Initial value of index when loop is first entered.
4. M_2 = terminal parameter	Final value of index, which when reached, causes exit from loop.
5. M_3 = incrementation parameter	Increment value added to the index variable each time the loop is traversed.

The range (or boundaries) of the DO loop is from the first executable statement following the DO statement to and including Statement N (where N is the programmer-assigned number of the last statement associated with the DO statement).

A FORTRAN DO statement example will summarize the discussion up to this point.

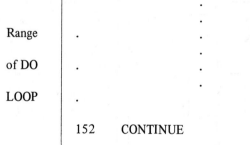

DO 152 I=1, 10, 1

Range of DO LOOP

. . . (First executable statement following DO statement)

152   CONTINUE

For the FORTRAN DO example, all statements within the loop (range) are to be repeated 10 times, the CONTINUE statement is a valid one and serves as a terminating statement that fulfills the technical requirements for the FORTRAN DO loop. This is not the only method to denote the loop boundary, but it is a simple one. As shown here, the DO statement contains all the necessary loop parameters.

Several forms of loop operation are shown in Figure 6.3 to illustrate the FORTRAN DO statement. The flowcharts are deliberately not labeled. The reader might as well begin to "sight read" flowcharts in a manner similar to that used by a musician who is handed a music score for the first time at a rehearsal. In Flowcharts (1) and (2), the loop parameters M_1, M_2, and M_3 are compressed into one symbol which is quite different from the ANSI flowchart symbol standard presentation. Flowcharts (3) and (4) are definitely related to FORTRAN DO statements. One of

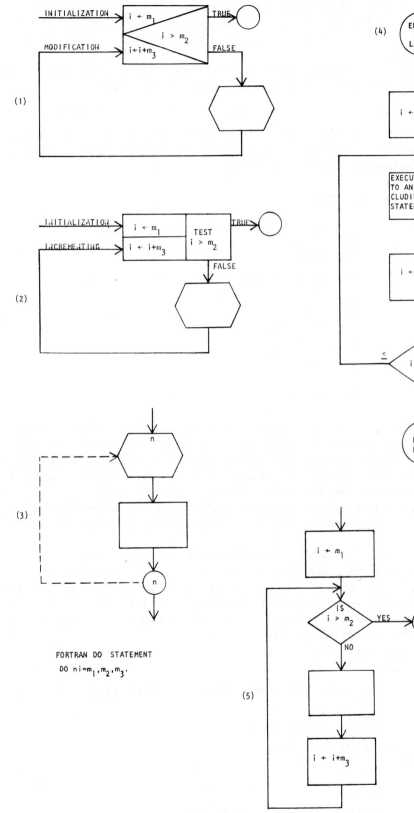

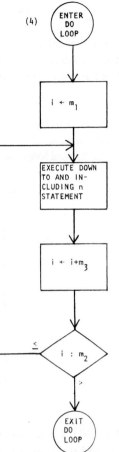

FIGURE 6.3   Representative DO Loops with Flowcharts

them shows the range of the DO statement by the end statement notation and the remaining flowchart leaves nothing to chance; the boundaries of the DO loop are cited. The ANSI form is the preferred one and accommodates a wide range of loop operations. The cross referencing between the flowchart and program statements was detailed in an earlier chapter.

The ALGOL equivalent of the FORTRAN DO statement[4] is shown in Figure 6.4 and a meta-language notation is introduced to the reader as a method of describing high level language statements.*

$$< \text{for statement} >::=<\text{for clause}><\text{statement} >|< \text{label} >:<\text{for statement} >$$

The meta-language notation shows self embedding and recursive properties; a definition defined in terms of itself. This will be pursued later. For the time being, the ALGOL reference[5] graphically presents the FOR statement semantics. A typical loop operation is illustrated here.

```
ALGOL FOR example:

          FOR  I: = 20  STEP 3 UNTIL 100   DO

        ┌ BEGIN

          S: = S+A [I];
                 ⋮
          STATEMENT;
                 ⋮
        └ END
  └── END
```

As noted here, the DO operation boundaries (delimiters) are BEGIN and END and the FOR boundary is the same word also, END. The particular block structures and indentation are characteristics of this language.

The execution of the FOR statement proceeds in the following manner: The control variable (I) is set equal to the initial expression (value being 20). If the value of the controlled variable is not greater than the final expression, the compound statement following the DO is executed (BEGIN . . . END). Otherwise, control is passed on to the statement following the end of the FOR loop. Each time the DO is performed, the indexing variable is incremented as specified (by the quantity three). Indentation serves only as a visual aid to the programmer and is not even necessary. ALGOL is free form.

The COBOL equivalent of the FORTRAN DO type of statement is flowcharted in Figure 6.5 together with its format. The PERFORM statement is used to depart from the normal ordered sequence to execute one or more procedures either a specified number of times or until a specified condition is satisfied. The PERFORM statement includes a means of return to the normal sequence. The example chosen is FORMAT 4 with one test condition present. The language permits numerous conditions to be imposed in one statement; a three-condition configuration is shown for illustrative purposes in Figure 6.6.[6,7]

The FORTRAN values are used here.

COBOL PERFORM example:

        PERFORM FORTRAN-ROUTINE VARYING I FROM
        1 BY 1 UNTIL I > 10
          GO TO NEXT-ROUTINE
      FORTRAN-ROUTINE

        .
        .
        .

      LAST STATEMENT OF THIS ROUTINE

---

*Although programming is not intended, the reader should be familiar with BNF (Backus Normal Form) and syntactical terminology in the software field.

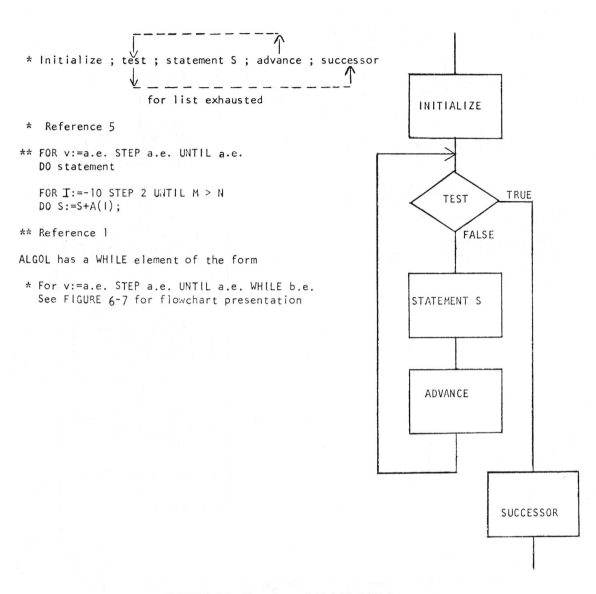

* Initialize ; test ; statement S ; advance ; successor

for list exhausted

* Reference 5

** FOR v:=a.e. STEP a.e. UNTIL a.e.
   DO statement

   FOR I:=-10 STEP 2 UNTIL M > N
   DO S:=S+A(I);

** Reference 1

ALGOL has a WHILE element of the form

* For v:=a.e. STEP a.e. UNTIL a.e. WHILE b.e.
   See FIGURE 6-7 for flowchart presentation

FIGURE 6.4 *Flowchart of ALGOL FOR Statement*

The execution of the PERFORM statement proceeds in the following manner. A loop indicator is set up or a counter is called FORTRAN-ROUTINE. The counter is initialized to the value of one and incremented by 1 for each loop cycle. The counter is initially tested and then tested on a per-loop cycle to see if the condition is met. If the condition is met, the procedure labelled FORTRAN-ROUTINE is executed. If the condition is not met, control is passed on to the statement following the PERFORM statement. In this case, it is the GO TO NEXT-ROUTINE. Here, the equivalent DO range is the called routine. When examining the COBOL specifications,[6.7] the PERFORM statement is flowchart illustrated.

The PL/1 language has two forms for the repetitive DO statements and they are flowcharted with their format in Figure 6.7. One initializes a control variable and repeatedly executes a group of statements as it increments this control variable and contains one condition test. The other repeatedly executes a group of statements as the first one with an additional specified condition (WHILE). All DO groups terminate with an END statement.

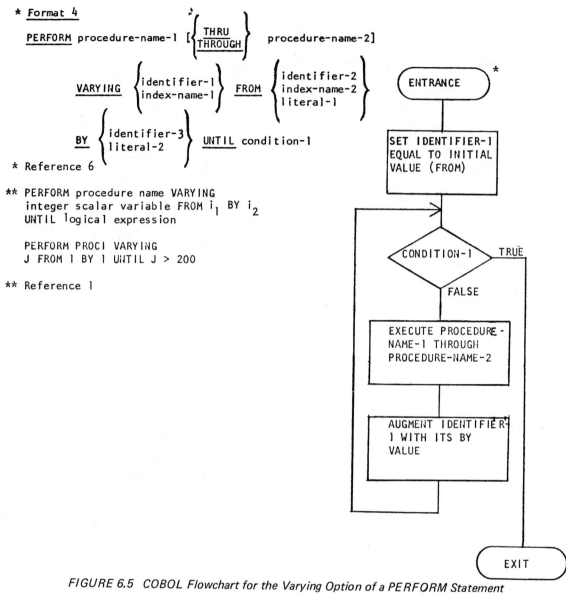

* Format 4

PERFORM procedure-name-1 [$\begin{Bmatrix} \text{THRU} \\ \text{THROUGH} \end{Bmatrix}$ procedure-name-2]

VARYING $\begin{Bmatrix} \text{identifier-1} \\ \text{index-name-1} \end{Bmatrix}$ FROM $\begin{Bmatrix} \text{identifier-2} \\ \text{index-name-2} \\ \text{literal-1} \end{Bmatrix}$

BY $\begin{Bmatrix} \text{identifier-3} \\ \text{literal-2} \end{Bmatrix}$ UNTIL condition-1

* Reference 6

** PERFORM procedure name VARYING
   integer scalar variable FROM i_1 BY i_2
   UNTIL logical expression

   PERFORM PROCI VARYING
   J FROM 1 BY 1 UNTIL J > 200

** Reference 1

FIGURE 6.5 COBOL Flowchart for the Varying Option of a PERFORM Statement
Having One Condition

PL/1 DO example:

LABEL	STATEMENT
L1:	DO I = 1 TO 10 BY 1;
	.
	.
	.
L2:	END:
NEXT:	...

It should be obvious to the reader when examining the flowchart Figures 6.3, 6.4, 6.5, and 6.7, that all the four languages contain: one loop; one condition; one initialization; and one incrementation. Program correctness, equivalence, and algorithm cataloging are done by using just the flowchart outlines. Illustrated here are several language type statements that in fact accomplish the same operation when flowcharted. It is rather easy to make this assertion by flowchart examples rather than examining each language statement by its syntax and for its

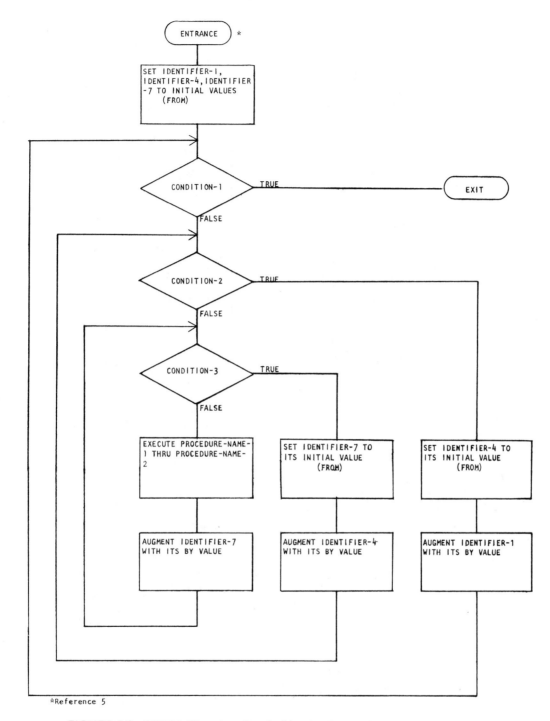

FIGURE 6.6 COBOL Flowchart for the Varying Option of a PERFORM Statement Having Three Conditions

semantics. For illustrative purposes, they are called DO type statements. Three out of four use the word DO. Also as noted in the illustrations, the parameters are initially tested prior to execution. This may not always be true when compiling the source input. The test may be performed after one execution of the loop.

Of the programming statements considered here, the DO statement alone has all the facilities to perform a loop operation. The remaining two in conjunction with each other can perform a loop operation. Before leaving the DO statement, we should mention some program restrictions which insure that the DO parameters are not invalidated. Each of the four language specifications should be consulted for special requirements and peculiarities.

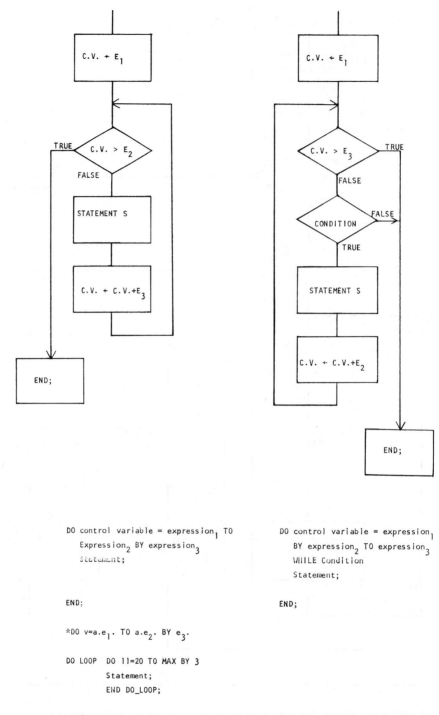

FIGURE 6.7 Flowchart of PL/1 DO Statement

Most of the restrictions are in relation to nested loop operations. The software restriction is necessary to insure a given loop is completed before exiting. Entrance into the loop must be at an entry point (initialization) to insure that the loop will actually be executed in its intended form.

GO TO STATEMENT

The GO TO statement is about the simplest in the programming language, but there are some who feel the GO TO statement is a major villain in programs that are difficult to understand and debug.[9]

The GO TO statement is directly related to the unconditional transfer instruction at the level of the machine code. It permits the programmer to alter arbitrarily the sequence in which the program statements are executed. The statement permits a forward transfer or a return (or reverse) transfer. The latter function permits return from a called routine or repetition of certain steps in a program (loop operation). In some cases, the GO TO may be thought of as a final statement of a preceding function or block terminal statement. The unconditional transfer statement which begins with the key words GO TO is essential in all FORTRAN programs. The GO TO examples are reviewed in FORTRAN. Analogies are drawn between the FORTRAN IV and PL-1 languages (Table 6.1). The use of the GO TO with a label variable as its object in PL/1 is analogous to the use of an assigned GO TO in FORTRAN. And, the computed GO TO in FORTRAN can be achieved in PL/1 by using a subscripted label variable.

For loop operations here, the GO TO can not stand alone. A single GO TO can create a loop operation without an exit (or escape).

STATEMENT NUMBER	STATEMENTS
LABEL 1	STATEMENT
.	
.	
.	
LABEL 3	GO TO LABEL 1

Exiting from this infinite loop can only be accomplished by a decision to branch by testing a modifiable condition that must be available within the body of the program statements.

The GO TO statement will be further illustrated in the IF statement section. For the moment, the GO TO flowcharting will be confined to FORTRAN IV, since this language has three types of GO TO statements:

1. Unconditional GO TO Statement
2. Assigned GO TO Statement
3. Computed GO TO Statement

The unconditional GO TO statement representation on a flowchart is not easily recognizable. A FORTRAN program is used to illustrate the forward and reverse transfers in Figure 6.8. As noted, cross referencing is used to demonstrate unconditional transfers.

The assigned GO TO statement has the following FORTRAN form:

$$\text{GO TO } i\,(k_1, k_2 \ldots, k_n)$$

where i is an integer variable reference, and the k's are statement labels. The integer variable can be assigned an arbitrary value during a sequence of statements and this may be performed frequently. At the time of execution of an assigned GO TO statement, the current value of the variable (i) is compared with one of the statement labels in the parenthesized list. The execution causes the statement identified by that statement label to be executed next. Thus, if i in the above format is equal to k_1, control goes to Statement k_1. Similarly, if $i = k_2$, control goes to k_2, and so on.

TABLE 6.1.   GO TO Statement

	FORTRAN IV[4]	ALGOL 60[5]	COBOL[6,7]	PL/I[8]
Unconditional Transfer	GO TO Statement	GO TO Expression	GO TO Procedure – Name	[GO TO] Label – [GOTO] Constant
Assigned Statement	GO TO i, $(k_1, k_2, \ldots, k_n)$			[GO TO] Label [GOTO]
Computed Statement	GO TO $(k_1, k_2, \ldots, k_n), i$		GO TO Procedure – Name DEPENDING ON Identifier	[GO TO] Element – [GOTO] Label – Variable (e.g., Subscripted Variable)
				GO TO Label (K)

FORWARD TRANSFER

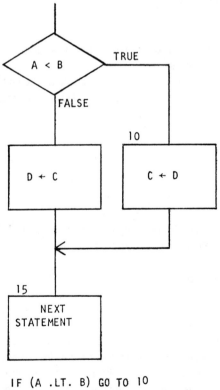

```
    IF (A .LT. B) GO TO 10
    D = C
    GO TO 15
10  C = D
15  NEXT STATEMENT
```

REVERSE TRANSFER

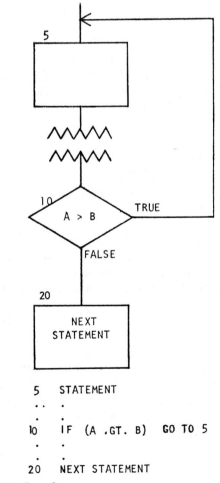

```
5   STATEMENT
 .. .
 .   .
10  IF (A .GT. B)   GO TO 5
 .   .
 .   .
20  NEXT STATEMENT
```

FIGURE 6.8   FORTRAN GO TO Transfers

EXAMPLE OF FORTRAN ASSIGNED GO TO STATEMENT

STATEMENT NUMBER	STATEMENT TEXT
	ASSIGN 20 to K
5	GO TO K, (10, 20, 30, 40)
10	STATEMENT
20	STATEMENT
25	A = 40
30	B = 30
34	C = A − B
50	ASSIGN C TO K
60	GO TO K, (10, 20, 30, 40)

In the illustration, the variable K has the value 20 initially. After the assigned GO TO is executed, the next statement to be executed is 20. Thus, a jump is made from Statement 5 to Statement 20. Later, the variable K has a computed value of 10. When Statement 60 is executed, a jump back to 10 is made. The assigned GO TO Statement is similar to the computed GO TO in that it allows a multiple transfer choice of control. However, the computed GO TO is quite useful when transfer of control can be made dependent on some code or key.

The computed GO TO statement may be used as a switch for branching to one of several places in the program depending upon the integer value at the time of testing:

$$\text{GO TO } (k_1, k_2, \ldots, k_n), i$$

where the k's are statement labels and i is an integer variable reference.

EXAMPLE OF COMPUTED GO TO STATEMENT IN FORTRAN

Statement No.	STATEMENT TEXT
1	. . . . (STATEMENT)
2	. . . . (STATEMENT)
3	. . . . (STATEMENT)
4	. . . . (STATEMENT)
5	GO TO (18, 20, 27, 52), K
6	. . . . (STATEMENT)
(ETC.)	(ETC.)

In this particular example, the GO TO depends on the value of integer variable K at the time of execution of the GO TO statement. If previous processing has made K = 1, the program transfers to Statement No. 18; if K = 2, the control transfers to Statement 20; if K = 3, to 27, and if K = 4, to 52. This example is flowcharted in Figure 6.9 where the alternate choices for transfer are shown as a series of decision symbols. Of the two, the branch table is more appropriate here, since the GO TO example demonstrated a switch operation having multiple branching options.

The intent here is to demonstrate flowcharting and to develop an understanding of GO TO operations for application in conjunction with IF statements. There are FORTRAN rules qualifying GO TO usage that should prevent any compiling errors. However, the problem of handling GO TO statements exists. The same criticism of GO TO FORTRAN IV statements is applicable to any language of GO TO utilization.[9,10,12]

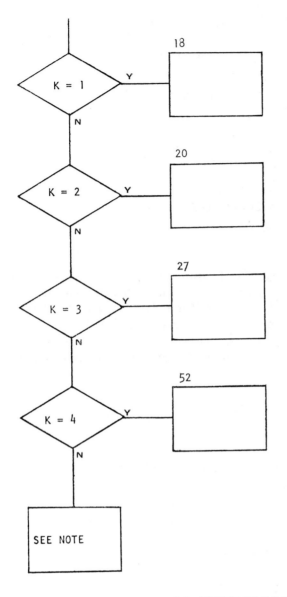

CODE K EQUAL TO	GO TO
<1	NEXT
1	18
2	20
3	27
4	52
>5	NEXT

GO TO (18,20,27,52), K

NOTE: THE MAXIMUM VALUE OF K
DEPENDS ON THE LANGUAGE PROCESSOR.
IN THE EVENT THE VARIABLE VALUE
EXCEEDS OR DOES NOT CORRELATE
WITH ONE OF THE STATEMENT
LABELS IN THE PARENTHESIZED
LIST, THE PROGRAM MAY TERMINATE
OR EXECUTE THE NEXT STATEMENT
DEPENDING ON THE LANGUAGE
PROCESSOR.

FIGURE 6.9   FORTRAN GO TO Conditional Transfers

IF STATEMENT

IF statements are examined in flowchart form, comparisons are made, and loop configurations are shown using GO TO statements. For purposes here, the IF statement is defined as a conditional statement that performs an operation upon occurrence of some condition. A simple condition is a single relation test of the following form:

IF (e) S or IF (e) THEN S

where e is an expression and S is an executable statement.*

In general, FORTRAN IV, ALGOL, COBOL and PL/1 possess this basic form. Syntactically, they may differ, but the semantics and the flowchart representation for a single condition are identical as shown in Figure 6.10. In operation, the IF condition is evaluated and if true, a branch is made. When the test is made and the result proves false (not true), the next statement is executed or the data path is continued and the branch is ignored. As

*In FORTRAN IV, e is a logical expression and S is any executable statement, except a DO statement or another logical IF statement. The syntax and semantic requirements of each language should be determined.[4,5,7,8]

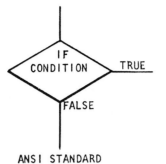
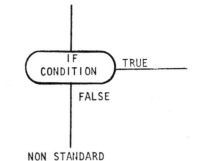

A. PRIMITIVE IF STATEMENT FORM.

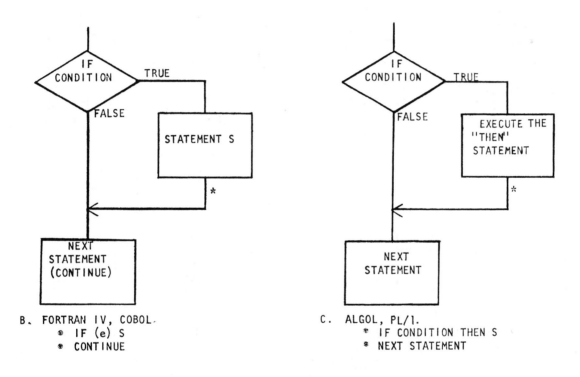

B. FORTRAN IV, COBOL.
   ⊛ IF (e) S
   ⊛ CONTINUE

C. ALGOL, PL/1.
   ⊛ IF CONDITION THEN S
   ⊛ NEXT STATEMENT

* THE NEXT STATEMENT IS SYMBOLIC. "NEXT" IMPLIES CONTIGUOUS.
THE TRUE PATH COULD HAVE A GO TO STATEMENT, AND "NEXT" WOULD
REFER TO THE FOLLOWING STATEMENT AND NOT NECESSARILY THE
NEXT CONTIGUOUS STATEMENT.

FIGURE 6.10 *The IF Statement as Executed in Four Languages*

shown here, the main path is vertical (false) and the alternate path is to the right (true). A tree can be constructed by substitution and concatenation using the primitive IF flowchart form to give a two way switching effect (Figure 6.11). The binary tree implementation is quite difficult to code considering the primitive IF statement allows only one branch. This is demonstrated in Figure 6.12 along with the IF–ELSE combination.

The ELSE form adds another dimension of control, since two alternate paths are available.

IF condition THEN statement-1, ELSE statement-2

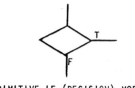

A. PRIMITIVE IF (DECISION) MODULE

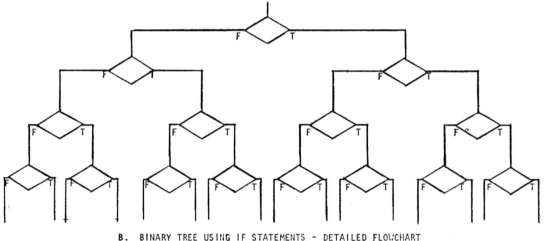

B. BINARY TREE USING IF STATEMENTS - DETAILED FLOWCHART

A REFERENCE TO A DETAILED   PRE-DEFINED PROCESS SYMBOL
REPRESENTATION

C. TOP LEVEL FLOWCHART OF PART B

FIGURE 6.11   A Binary Tree Implementation of IF (Conditional) Statements

The IF–ELSE combination is effectively a two-way switch. A comparison of two-way switching using the IF statement with and without the ELSE is shown in Figure 6.12. The ELSE form states that the results of the test are either true or everything is assigned to the false state. To obtain the same results (two selections) using the basic IF form (Figure 6.12), the NEXT STATEMENT of the second IF level would never occur (null set). Also, the NEXT STATEMENT of the ELSE can be used to merge the two data paths of both branches.

The ELSE will now be examined in the expanded form. As in the multiple branching of the GO TO, the ELSE form can be expanded to govern execution of any number of alternate selections, rather than just two.* The flowchart in Figure 6.13 illustrates the constructions in COBOL and PL/1, but the coding is in PL/1. The same can be said for ALGOL, but for purposes here, the block form presentation is introduced and used to depict nesting.

---

*This facility is not available in FORTRAN IV per se, but it can be constructed from basic elements and compiled to simulate or emulate a multiple decision configuration.

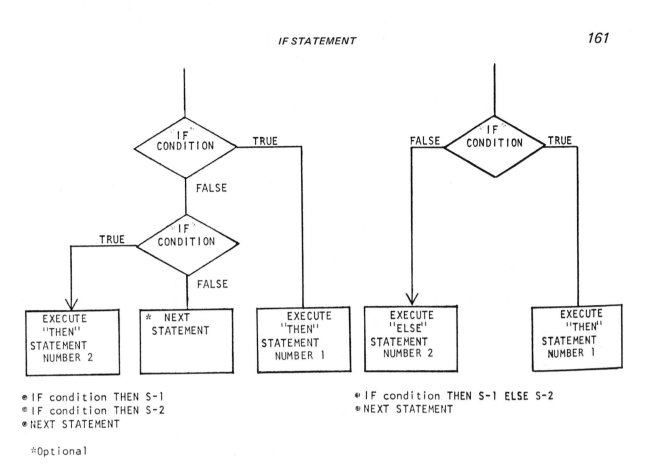

● IF condition THEN S-1
● IF condition THEN S-2
● NEXT STATEMENT

● IF condition **THEN** S-1 **ELSE** S-2
● NEXT STATEMENT

*Optional

FIGURE 6.12   Two-way Selection Using IF Statements

IF example in PL/1 language:

> IF condition-1 THEN statement-1;
> ELSE IF condition-2 THEN statement-2;
>   ELSE
>
>   .
>   .
>   .
>
>   ELSE IF condition-N THEN statement-N;
>   ELSE statement-M;
> NEXT STATEMENT

When the IF statement above is executed, the following action is taken:

1. If the result of condition-1 is true, statement-1 is executed and control then passes implicitly to the NEXT STATEMENT.

2. If the condition is false, the ELSE statement is executed. For purposes here, the ELSE statement contains an IF statement. In this case, the IF statement is said to be nested.

3. IF statements within IF statements are shown in indentation form where an ELSE statement is associated with a previous IF condition. Going from left to right any ELSE that is encountered is considered to apply to the immediately preceding IF that has not been already paired with an ELSE. See block form in Figure 6.13.

4. Each condition is tested to find if it is true or false by repeating above items 1, 2, 3 until all tests have been executed or exhausted. If all prove to be false, control passes to the statement-M and then NEXT STATEMENT.

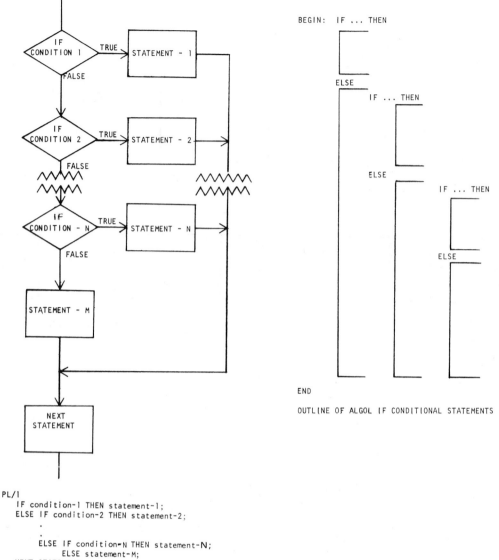

BEGIN: IF ... THEN

ELSE

            IF ... THEN

            ELSE

                        IF ... THEN

                        ELSE

END

OUTLINE OF ALGOL IF CONDITIONAL STATEMENTS

PL/1
    IF condition-1 THEN statement-1;
    ELSE IF condition-2 THEN statement-2;
            .
            .
            ELSE IF condition-N THEN statement-N;
                ELSE statement-M;
    NEXT STATEMENT

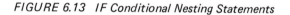

FIGURE 6.13 IF Conditional Nesting Statements

If the last IF statement did not contain an ELSE, and condition-N is false, statement-M is skipped and control passes to the NEXT STATEMENT.

Although FORTRAN IV does not include IF nesting, it has a three way switch not found in ALGOL, COBOL, or PL/1. FORTRAN IV has an arithmetic IF statement using the form

$$\text{IF (e) } k_1, k_2, k_3$$

where e is any arithmetic expression and the k's are statement labels.

Three kinds of values may be sensed in examining e. If e is found to be negative, the program goes to Statement k_1; if e is zero, the program goes to Statement k_2; and if e is positive, the program goes to k_3.

The flowchart equivalent of the arithmetic IF statement is shown below. (Two alternative representations are shown.)

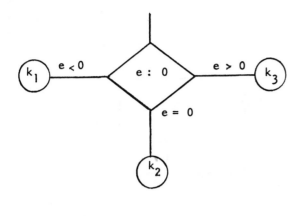

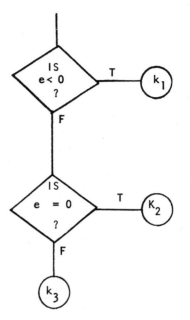

ARITHMETIC IF

ARITHMETIC IF
(ALTERNATE REPRESENTATION)

The conditional (IF) transfer permits a branch away from the current flow path. To obtain a loop capability, it is necessary to combine or pair an IF with a GO TO. The location of the IF with respect to the GO TO permits a test to be made on loop entry or after one processing cycle has been completed. Both are presented using FORTRAN statements in Figure 6.14. Testing on loop entry requires a false test to maintain loop operation. Testing prior to loop cycling requires a true test result to maintain loop operation.

The combination of IF and GO TO has considerable flexibility and versatility to accommodate a wide variety of programming operations.[11]

RECURSIVE OPERATIONS

In trying to learn the concept of recursion,* one frequently has trouble in distinguishing recursion from iteration. In most cases and in available languages (FORTRAN and COBOL), an operation is most efficiently performed iteratively, even when there are advantages in defining it recursively. The penalty paid in many recursive systems is that the automatic (software or hardware) mechanism is brought into play even when the program is nonrecursive, with a consequent waste of time. There are various ways around this and the programmer should indicate whether or not his functions or procedures are recursive.

Iteration and Recursion

Iteration is the repeated operation of a statement set until some condition is met, each statement set being carried to completion, the condition being examined, and a new performance commenced if the test condition is met. Once the loop operation has begun, the iteration is always successfully completed. A finite set of loop parameters, each of which is finite in value, assures this.

In contrast to this, a recursive operation always has the possibility of not terminating and of being infinitely regressive. Some recursive operations may terminate when given certain inputs, but may not terminate for others. It is theoretically impossible to determine whether a definition will terminate in the general case, although it is often possible to show that particular cases will or will not terminate.[18-21] Therefore, any recursive definition

*Recursion is the name given to the process of defining a function or relationship in terms of itself.[13]

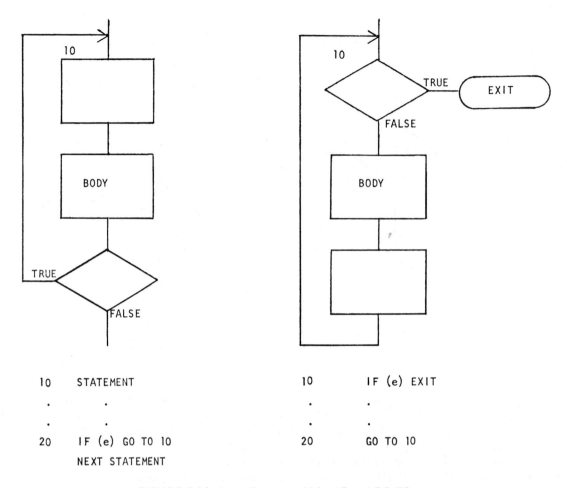

10	STATEMENT		10	IF (e) EXIT
.	.		.	.
.	.		.	.
20	IF (e) GO TO 10		20	GO TO 10
	NEXT STATEMENT			

FIGURE 6.14   Loop Construct Using IF and GO TO

must always contain one nonrecursive alternative to avoid a circular path (infinite loop). For this sort of definition, the terminal (exit) condition is the initial test condition:

$$f(x) = \begin{cases} \text{IF } P_1, \text{ THEN } e_1 \\ \\ \text{ELSE, IF } P_2 \text{ THEN } e_2 \end{cases}$$

where P_1 is the initial test (a Boolean value, TRUE, is required).

e_1 is the exit or return — an expression having a conditional provision[22] in the form

$(P_1 \rightarrow e_1, P_2 \rightarrow e_2, \ldots, P_n \rightarrow e_n)$

Also, recursion can occur indirectly through a chain of function definitions that eventually return to the original ones.[14,23] With this brief introduction, we shall examine some aspects of recursive operations in flowchart form. The interested reader is encouraged to pursue the subject using the cited references.

Nesting

Recursion and iteration are examined in terms of operation and language form. In both cases the flowchart will exhibit loops within loops, where the range of the inner loop must be enclosed by the range of the outer one. A nested loop is depicted in the diagram.

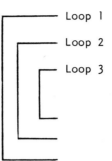

Loop 1

Loop 2

Loop 3

For purposes here the common form of iteration involves a known number of repetitions. Such iterations are usually programmed as a simple count using DO type statements. A loop within a loop is obtained by using subscripted variables in an array notation. A typical nested loop operation is shown in Figure 6.15. It uses FORTRAN program DO loops to scan Array (I, J, K) in Figure 6.1. The cube is scanned element by element, left to right, and then row by row like a television screen. And then from front to back for the third dimension so that each element in the cube has been accessed. In a volume memory, this is akin to scan, sweep, and search operations for sort and merge functions.

Examination of Figure 6.15 shows the conditional transfer is a two way switch, and the indices are finite, thus ensuring against infinite loop operation. The three sequential CONTINUES in the program listing (part of Figure 6.15) are not necessary when one can suffice. They are included so that the nested loop configuration is not obscure to the beginner. Nesting of DO loops is not limited to three loops. The particular compiler (not language) determines the maximum allowed depth of nesting.

The recursive nested loop avoids explicit counting (indexing). Control will remain in the subroutine until a matching test condition occurs and escape is accomplished. The recursive operation intrinsically needs the concept of a conditional expression, and the IF–THEN–ELSE format provides this. The decision symbol provides a three way switch (Figure 6.16) to accomplish a stacking concept in the manner shown.

SWITCH POSITION	OPERATION	DESCRIPTION
1	Decrement/predecessor	Operation is completed, decreasing the depth level by one; if last (top) level, the stack becomes empty.
2	Loop	Operation continues to completion unless interrupted.
3	Increment/successor	Current operation is interrupted and an addition to the stack is made, increasing the depth level by one.

In order to relate and compare a recursive nested loop with the DO type nested loop, a three level stack presentation is shown in Figure 6.17. To be consistent, the first decision symbol in each representation is not recursive, but is an initial test as discussed earlier. Obviously, since this is the only exit path, all subsequent operations when completed must return for the final test to exit from the recursive routine. Either each subroutine has access to the exit test (nested three level) or by decrementing one level at a time until the decision symbol with the exit is involved. The former is random and the latter is sequential.

In the recursive presentation, each subroutine is permitted to operate in a loop fashion independent of its adjacent environment. One may ask "When is the procedure function completed?" And the response is "That's a good question." Unlike the nested iteration, no finite count or indexing is shown. From a visual presentation, when the iteration is completed, exit is from the bottom of the flowchart, because the predicted operation has been performed. On the other hand, the recursive operation returns to the initial nonrecursive decision symbol to exit. Even on exiting, one can not be positive that the procedure was executed correctly as shown here.

Recursive procedures are particularly useful when they allow a program to be formulated in a simple manner,[15] without the use of multiple DO loops, or when the resulting storage requirements[16] are smaller than would result from the equivalent use of nonrecursive procedures. However, in many instances use of recursive procedures requires a longer execution time[17] than does the use of an equivalent program without recursive procedures.

The reader is cautioned not to mix recursive operation indiscriminately with language, compiler, and computer; software and hardware. Also, a computer installation may have FORTRAN, COBOL, ALGOL, PL/1, LISP, and APL processors available on the same computer. The software recursive mechanism such as a stack concept[24] may

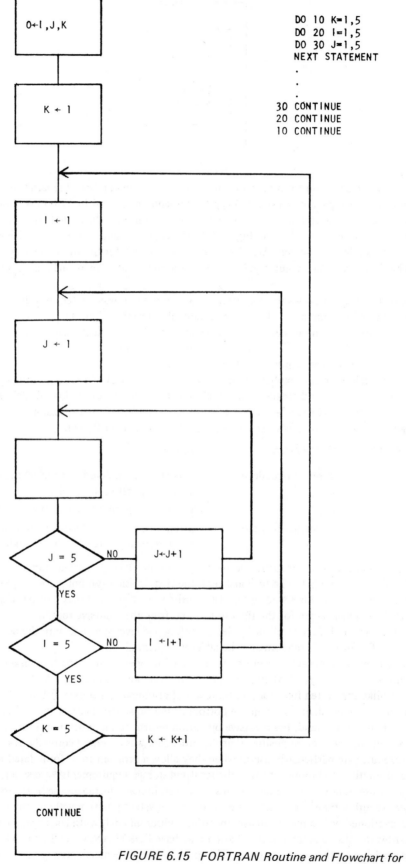

```
DO 10 K=1,5
DO 20 I=1,5
DO 30 J=1,5
NEXT STATEMENT
    .
    .
    .
30 CONTINUE
20 CONTINUE
10 CONTINUE
```

FIGURE 6.15   *FORTRAN Routine and Flowchart for Complete Scan of Array (I, J, K) in Figure 6.1*

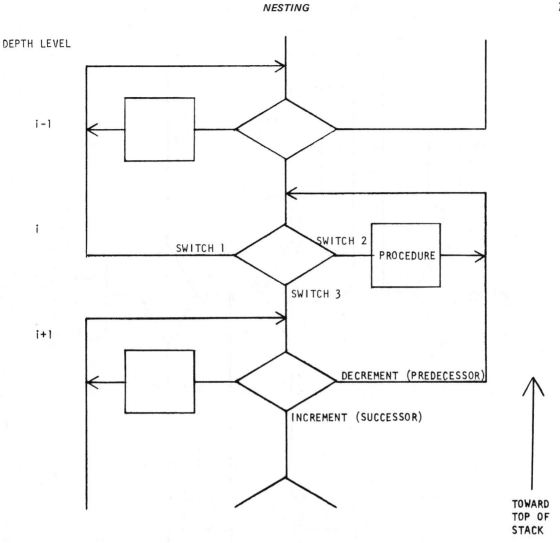

DEPTH LEVEL

i-1

i

SWITCH 1

SWITCH 2

PROCEDURE

SWITCH 3

i+1

DECREMENT (PREDECESSOR)

INCREMENT (SUCCESSOR)

TOWARD
TOP OF
STACK

FIGURE 6.16   Flowchart Push-down Stack Representation

be used to compile source statements, yet the high level language is nonrecursive. On the other hand, the stack concept may be hardware implemented as provided in the Burroughs B5000 and B6000 series. The distinction between a recursive language such as ALGOL, PL/1, LISP and APL, and a nonrecursive language such as FORTRAN is that in FORTRAN the programmer[25,26] who wishes to use recursion has to set up the means for himself, and may be unable to do so, whereas in a recursive language, the mechanism is provided with the system. The IF statements in COBOL were shown nested. Nest calls of subroutines are possible subject to elaborate rules which, among other things, rule out recursive calls.

Before leaving the subject of nesting, we present Figure 6.18, which summarizes the loop formation rules presented in an earlier chapter. The rules are illustrated in terms of DO type statements. The nested DOs in Figure 6.18B indicate legal (solid arrow) and illegal (broken arrow) control transfers.

Recursion Examples

Many functions are conveniently defined recursively. Perhaps the best known example of a recursively defined function is the factorial function for a positive integer. The factorial of a number is the product of all the integers from 1 up to and including the number, and expressed mathematically as follows:

$$\text{FACTORIAL } N = N! = \prod_{i=1}^{n} i$$

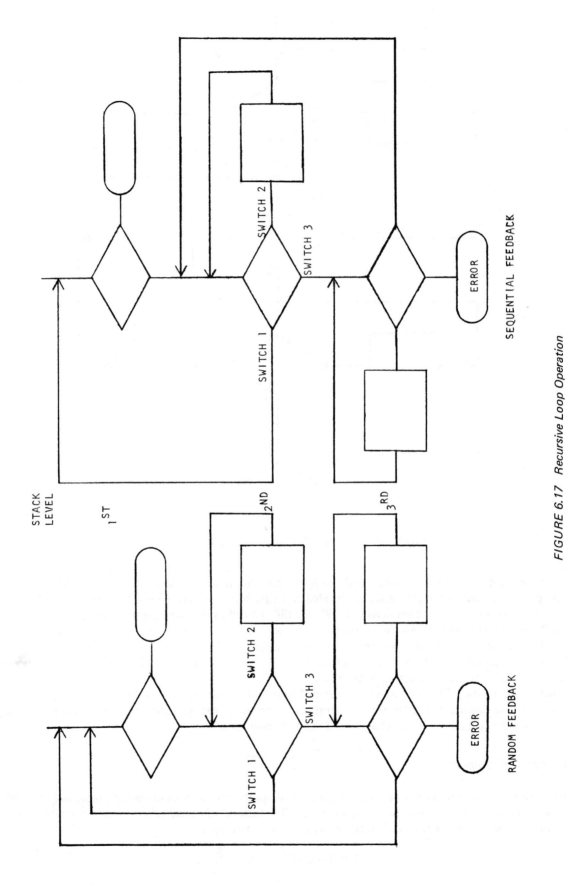

FIGURE 6.17 Recursive Loop Operation

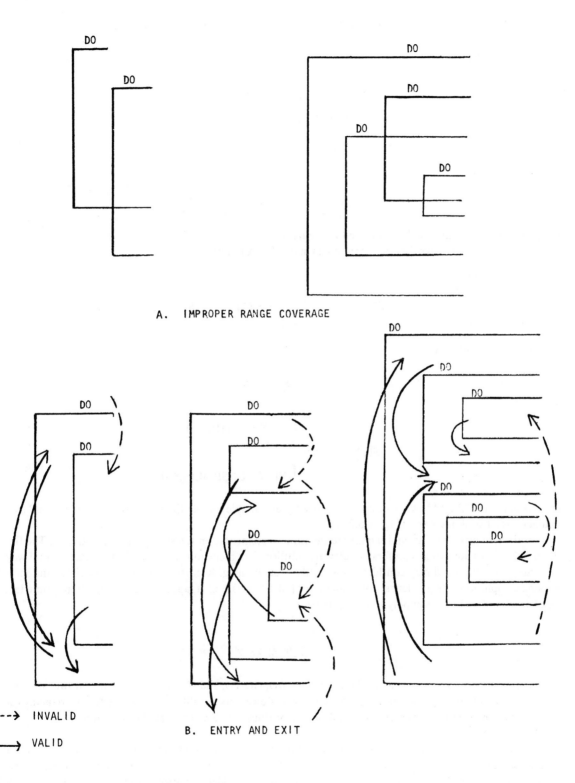

FIGURE 6.18 DO Loop Nesting Rules

and computed as follows:

$$N! = (N)(N-1)(N-2)\ldots(2)(1) = (1)(2)(3)\ldots(N-1)(N)$$

and recursively expressed

$$f(n) = \begin{cases} \text{IF } n = 0 \text{ THEN } 1 \\ \\ \text{ELSE } n \cdot f(n-1) \end{cases}$$

or (from Reference 22)

$$N! = \left(n = 0 \to 1, T \to n \cdot (n-1)! \right)$$

The equations are detailed and flowcharted in Figure 6.19.

Another typical recursive function is the SUM and SIGMA function.

Iteration Loop

$$\text{SUM} = \sum_{i=1}^{n} S_i = S(1) + S(2) \cdot \ldots + S(N)$$

Recursive Definition

$$\sum_{i=1}^{n} = S_n + \sum_{i=1}^{n-1} S_i$$

$$\text{SIGMA}(X) = \begin{cases} \text{IF } X = 0 \text{ THEN } 0 \\ \\ \text{ELSE } X + \text{SIGMA}(X-1) \end{cases}$$

The example in FORTRAN is a vector (single subscripted variable), while the recursive example in LISP treats the argument as a list of numerics. (Figure 6.20)

In both examples (factorial and sigma) the iteration implementation is a single subscript variable. The flowchart presentation does not reveal a preferable software solution. Yet recursion is frequently the most natural way of expressing a function and may considerably simplify programming. Recursive programs are written and include numerical applications (e.g., highest common factor (HCF), square root, approximate integration, prime factor calculation, and evaluation of polynomial expression).

SUMMARY

Four high level language statements of the operations DO, GO TO, and IF are translated into flowchart form. They are examined and shown to be equivalent, because their flowcharts are alike.[27] Programs today are being cataloged and classified in terms of flowchart configuration. Algorithms are being translated into flowcharts for analysis and solvability.[18-20,28]

Recursive operations were introduced because of their nesting configuration. The subject was pursued further since considerable interest has been shown and witnessed by an extensive bibliography on the subject (12 references here). Recursion is widely regarded at present as an interesting academic pursuit. When the number of examples and applications is tabulated, recursive techniques are found to be another tool for the programmer, so they are included here. The mechanism (stacked concept) for recursion is quite simple in terms of both software and hardware.

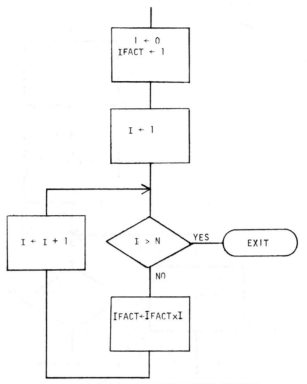

THREE ITERATIVE SOLUTIONS USING FORTRAN

(ALTERNATIVE NO. 1)
IFACT = 1
DO 20 I = 1, N
IFACT = IFACT × I
20 CONTINUE

(ALTERNATIVE NO. 2)
IFACT = 1
I = 1
10 IF (I - N) GO TO 20

I = I + 1

IFACT = IFACT × I
GO TO 10
20 CONTINUE

(ALTERNATIVE NO. 3)
IFACT = 1
K = N
IF (K) 20, 20, 10

10 IFACT = IFACT × K

K = K - 1
GO TO 10
20 CONTINUE

RECURSIVE PROCEDURES AND CALLS

ALGOL

INTEGER PROCEDURE IFACT (N),
VALUE N; INTEGER N;

IFACT: = IF N = 1, THEN 1

        ELSE N×IFACT (N-1);

PL/1

IFACT: PROCEDURE (N) RECURSIVE;

IF N = 1 THEN RETURN (1);

ELSE RETURN (N*IFACT(N-1))

END IFACT,

LISP[29]

DEFINE ((

(FACTORIAL (LAMBDA (N) (COND ((ZEROP N) 1)
  (F (TIMES N (FACTORIAL (SUBI N))))))))

))

THE FLOWCHART ABOVE CORRESPONDS TO FORTRAN ALTERNATIVE NO. 1 FOR ILLUSTRATIVE PURPOSES.

FIGURE 6.19 *Factorial Computation*

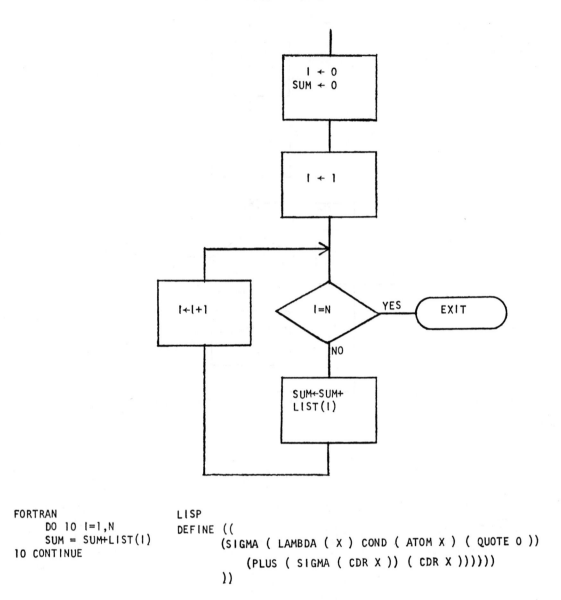

FORTRAN
```
FORTRAN
      DO 10 I=1,N
      SUM = SUM+LIST(I)
10 CONTINUE
```

LISP
```
LISP
DEFINE ((
      (SIGMA ( LAMBDA ( X ) COND ( ATOM X ) ( QUOTE 0 ))
          (PLUS ( SIGMA ( CDR X )) ( CDR X ))))))
      ))
```

FIGURE 6.20   Recursive Summation Comparison with FORTRAN

The problems included at the end of this chapter require more than just flowcharting. The reader is encouraged to become familiar with the four high level languages presented. Further familiarity at the machine level can be obtained by mastering an assembler language. Recursive programming can be obtained using ALGOL, PL/1, and LISP. However, the parenthesis format of LISP seems to act as a mental deterrent to its wide application.

REFERENCES

1. Cardenas, A. F., Presser, L., and Marin, M. A., *Computer Science,* Table 3, Expressions and Executable Statements. New York: John Wiley, 1972. DO pp. 339, GO TO pp. 337, and IF pp. 338.

2. IBID, pp. 324-355.

3. Katzan, Jr., Harry, *ADVANCED PROGRAMMING – Programming and Operating Systems,* pp. 205-207. Van Nostrand Reinhold Company, 1970.

4. ANSI-X3, 9-1966 FORTRAN. DO pp. 14-15, GO TO pp. 13, and IF pp. 13.

5. *Report on the Algorithmic Language ALGOL 60,* Comm. ACM, Vol. 3, No. 5, pp. 299-314. May 1960 with typographical corrections as of April 1, 1962. FOR pp. 307-308, GO TO pp. 307, and IF pp. 307.

6. "Codasyl COBOL Journal of Development 1969," *Report 110-GP-1a.* Canadian Government Specifications Board, April 1970. PERFORM III 7-62 through 69, GO TO III 7-48 and 49, and IF III 7-51 and 52.

7. ANSI-X3.23-1968 COBOL. PERFORM pp. 2-93 through 98, GO TO pp. 2-87 and 88, and IF pp. 2-88 and 89.

8. *IBM System/360 Operating System PL/1 (F) Language Reference Manual,* IBM Form GC28-8201-3, June 1970. DO pp. 69-70, 373-376, GO TO pp. 68, 379-380, IF pp. 69, 380.

9. Wulf, William A., "Programming Without the GOTO," pp. 408-413, Ashcroft, Edward, and Manna, Zohar, "The Translation of 'GO TO' Programs to 'While' Programs," pp. 250-255. *Proceeding of the IFIP Congress 71,* Vol. 1. C. V. Freiman, ed. North-Holland Publishing Company, 1972.

10. Dijkstra, Edsger W., "GO TO Statement Considered Harmful," *Comm. ACM,* Vol. II, No. 3, pp. 147-148. March 1968.

11. Organick, Elliott I., *A FORTRAN IV PRIMER*, pp. 55-57, pp. 63-64. Addison-Wesley Publishing Co., Inc., 1966.

12. Manna, Zohar, and Vuillemin, Jean. "Fixpoint Approach to the Theory of Computation," *Comm. ACM,* Vol. 15, No. 7, pp. 528-541. July 1972.

13. Barrow, D. W., *Recursive Techniques in Programming,* pp. 1. London: Macdonald, 1968.

14. IBID, pp. 45

15. IBID, pp. 10-11

16. IBID, pp. 15

17. IBID, pp. 11

18. Strong, H. R., "Flowchartable Recursive Specifications," *Algorithm Specification,* Randall Rustin, ed., pp. 82-96, published by Prentice Hall, 1972.

19. Strong, H. R., "Translating Recursion Equations into Flow Charts," *Proc. 2nd. Annual ACM Symposium on Theory of Computing,* pp. 184-197. Northampton, Mass., May 1970.

20. Strong, H. R., "Translating Recursion Equations into Flow Charts," *Jr. Computer & Systems Science,* Vol. 5, No. 3, pp. 254-285. June 1971.

21. Green, M. W.; Elspas, B.; and Levitt, K. N. *Translation of Recursive Schemas into Label Stack Flowchart Schemas,* 5 pages. Menlo Park, Calif.: Stanford Research Institute, June 1971.

22. McCarthy, John, "A Basis For a Mathematical Theory of Computation," *Computer Programming and Formal Systems,* by Braffort, P. and Hirschberg, D., pp. 33-70. Amsterdam: North-Holland Publishing Company, 1963.

23. Wegner, Peter, *Programming Languages, Information Structures, and Machine Organizations,* pp. 228-291. McGraw-Hill Book Company, 1968.

24. Bycer, Bernard B. *Computer Systems Aspects of Magnetic Recording on Moving Media,"* Seminar on Advances in Magnetic Recording Engineering, April 24, 1972, Proceeding sponsored by Philadelphia Section, Committee on Basic Sciences, IEEE, pp. 8.1-8.49.

25. Morris, John, "Programming Recursive Functions in FORTRAN," *Software Age,* Vol. 3, No. 1, pp. 38, 40-42, January 1969.

26. Kugel, Herbert C. "An Experimental Approach to Recursive FORTRAN Subroutine Programming Under Operating System 360," *Software Age,* Vol. 3, No. 12, pp. 14-17. December 1969.

27. Ershov, A. P. "Theory of Program Schemata," *Proceeding of the IFIP Congress 71,* Vol. 1, pp. 28-45, C. V. Freiman, ed. North-Holland Publishing Company, 1972.

28. Bruno, John, and Steiglitz, Kenneth. "The Expression of Algorithm by Charts," *Algorithm Specification* edited by Randall Rustin, ed., pp. 97-115. Prentice-Hall, 1972.

29. Weissman, Clark, *LISP 1.5 PRIMER*, Belmont, Calif.: Dickenson Publishing Company, Inc., 1967.

PROBLEMS

1. Flowchart and write the DO loop expressions for the following scan pattern.

TOP TO BOTTOM

LEFT TO RIGHT

ANSWER

Same flowchart form as Figure 5.15 with the i and j indices interchanged (i ←j, j←i)

2. Flowchart the COBOL format

IF (condition-1) $\underline{AND}$ (condition-2) . . . [imperative statement(s)]

$$\left\{ \begin{bmatrix} \text{ELSE} \\ \underline{\text{OTHERWISE}} \end{bmatrix} \text{imperative statement(s)} \right\}$$

ANSWER

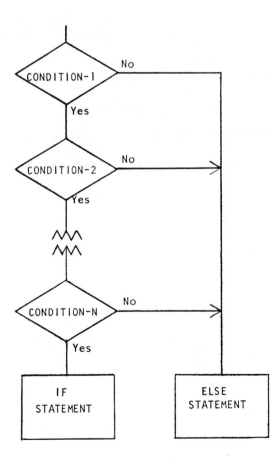

3. Write a single FORTRAN statement to code the following steps: (Hint — see Problem 2)

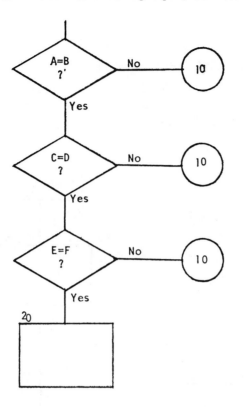

ANSWER

   IF (A .EQ. B .AND. C .EQ. D .AND.
      E .EQ. F) GO TO 20

10 STATEMENT

    .
    .

20 STATEMENT

4. Code the following routine in ALGOL. (Hint: IF recursive)

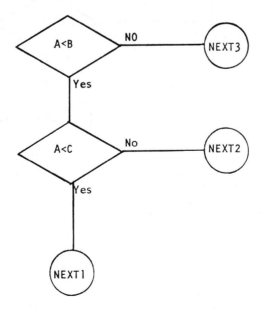

ANSWER
IF A ≥ B THEN NEXT3
ELSE IF A ≥ C THEN NEXT2
      ELSE NEXT1

5. Flowchart the PL/1 DO... TO... BY... END statement.

DO 1=7 to 15 BY 2;
    Statement Sequence
END;

ANSWER

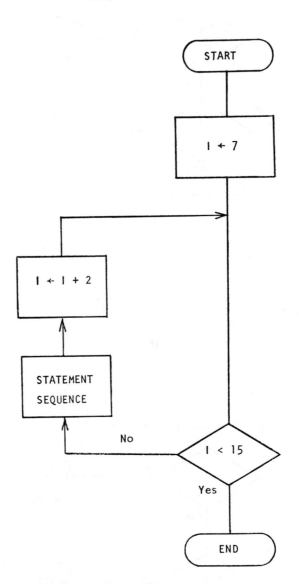

6. Flowchart the Fibonacci series: 0, 1, 1, 2, 3, 5, 8, . . . (Hint: see page 98 in *Designing Computer and Digital Systems* by C. G. Bell; J. Grason; and A. Newell, PUB. BY DIGITAL EQUIPMENT CORPORATION.)

ANSWER

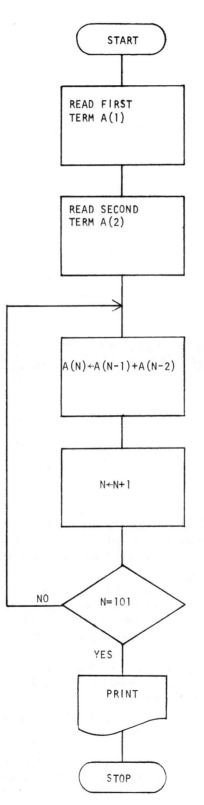

FIBONACCI SERIES

7

FLOWCHART WORK PROBLEM EXAMPLES

Problem solving utilizing flowcharts is demonstrated here by means of detailed work problem examples. The solution to each problem is flowcharted, analyzed, modified, and finalized. In each case the flowchart is carefully scrutinized and the logic paths are analyzed to determine if there is a better solution than the one at hand. Symbols are moved and rearranged to obtain patterns which modularize the program as much as possible. Every attempt is made to achieve coding simplification or to reduce the total effort by the programmer. When every method has been exhausted to reduce the complexity of the symbol configuration and flowchart, the text material within each flowchart symbol is then challenged. Carefully, the text material (operation or process) within each symbol is reviewed for redundance (repetition), relevancy (requirement), and necessity (performed by another process).

For example, in the same sequence just given, inclusion of clearing operations (setting to zero) can be found to be an unnecessary habit. Many storing operations are self-erasing, so reusing a temporary storage area need not always entail clearing its contents to zero. Writing and reading just those locations needed for a processing operation using an indexing scheme (block transfer number) avoids clearing a complete storage area for a small number of contiguous locations. In some cases, written programs are so general in their purpose that they are unduly complex and are not relevant to the solution of the problem.

Redundance is a problem which sometimes can be quite obscure. For example, a flag may be used to denote a given operation to be executed or not executed, and the flag is cleared (or turned off) for the opposite state. The result can be a routine such as that diagrammed.

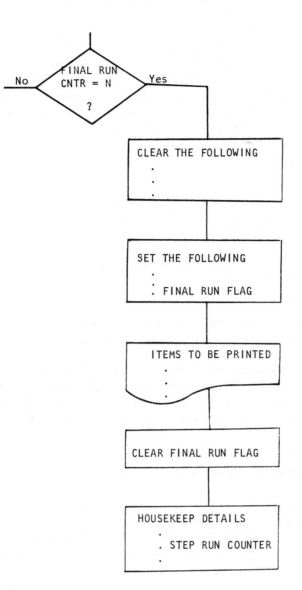

Here, the final run flag is set and interrogated for a printout operation. Once the printout has been accomplished, the flag is reset. Entrance to this logic path is via a final run. The printout is always accomplished on a final run basis and inline coding will do the printout operation, so there is no need for the flag.

The selected work examples given in this chapter form two categories. The first category is relevant to programming and coding in several high level languages (FORTRAN, COBOL, ALGOL, and PL/1) and processing equipment (computers). The work problem examples review several aspects that include the following:

1. Flowcharting
2. Programming
3. High Level Languages
4. Computer Configurations

The second category of work examples is devoted to software system design. The flowchart delineates equipment operation (hardware) by use of software (programming). The actual operation is spelled out in terms that are comprehensible to the programmer and form the basis of a software specification.

In a typical software project, the system design (as represented by the second category of examples in this chapter) is performed prior to the programming and coding (first category of examples). The reverse order of presentation is deliberately used in this chapter for simplicity.

The flowchart presentation procedure is a simple one. As a rule each flowchart symbol is identified and its operation is described in detail relevant to its function and within the total flowchart configuration. The exercises in this chapter maintain the above theme. The reader is encouraged to question not only the flowchart, but also the arrangement of the symbols within the flowchart. Experience has shown that a simplified flowchart solution will generally result in a less complex computer program. When asked to write a computer program, the programmer should begin by setting down a flowchart for the program. Then he should proceed to seek improvements and simplifications in the flowchart prior to writing the computer program.

COMPUTER PROGRAMS

In examples shown in this chapter, the procedure for problem solving consists of the following steps:

1. Understand the problem or requirement.
2. Draw a flowchart for the problem.
3. Code symbolically (label) each box of the flowchart.
4. Assign addresses from the flowchart.
5. Write out the final details.

The flowchart presents an overall visual picture of the problem, allowing one to organize the programming effort so that it can be handled one box at a time. Programming or coding can be generated from detailed flowcharts. With detailed charts, the programmer can think through the logical flow of steps before having to concern himself with addressing, constants, names, and the like. In no way is the programmer restricted, confined, or coerced to follow a specific procedure. His freedom for creativity and originality is not hindered and he can take full advantage of the high level language (compiler) and the host computer characteristics. He can rearrange his logic paths to obtain the desired effects subject only to the task requirements. The flowchart is his expression of a problem solution.

AREA UNDER A SINE WAVE — FORTRAN

The first example is very simple and can be run on an installation having FORTRAN facilities. A variety of flowchart configurations and coding solutions are possible.

The area under a sine wave is computed using a calculus procedure. The problem is to find the area under the sine wave from $X = 0$ to $X = \pi/2$. The area under the curve is computed as a summation of a contiguous series of rectangles using the rectangle area formula.

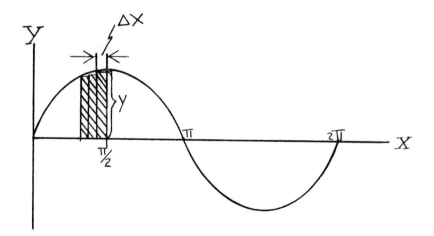

AREA (of each small rectangle) $= \triangle X \cdot Y$

where $\triangle X = \dfrac{\triangle \theta \cdot \pi}{180}$ in radians ($\triangle \theta$ in degrees)

$Y = SIN\ \theta$ for $0 \leqq \theta \leqq 90 = \dfrac{\pi}{2}$

The area is divided into thin strips each $\triangle X$ in width. The finer these divisions of X, the more accurate the approximation for the area. In this example, the X value is divided into 100 equal increments, so that

$$\triangle X = \left(\frac{1}{100}\right)\left(\frac{\pi}{2}\right)$$

The call routine for the sine function is assumed rather than computing the sine value each time using a polynomial expression.

The FORTRAN program for Figure 7.1 is given below:

```
10     START*
20     X = 0
30     T = 0
40     IF   (X .GT. 3.1416/2)  GO TO 90
50     X = X + (3.1416/2)/100
60     A = SIN(X)*((3.1416/2)/100)
70     T = T + A
80     GO TO 40
90     WRITE (6, 100) X, T
100    FORMAT (F10.8,4X,F10.8)
110    STOP
```

Examination of both the program and the flowchart (Figure 7.1) discloses alternate methods. Some flowchart rearrangements can be devised in response to a series of questions. Example: would reversal in the sequence of flowchart symbols 2 and 3 have any effect on the computation? Suppose the decision symbol were located after the symbol 5 (so that the symbols appear in the order 3, 4, 5, and 2); would it alter or disqualify the loop operation? The answer is NO in both cases.

Where there is an IF and GO set configuration as shown here, the FORTRAN DO statement method can be substituted. The constant $((\pi/2)/100)$ is computed each time according to the flowchart shown here. Computation time can be saved if the constant is computed once, stored, and given a label so the value can be retrieved whenever needed.

FORTRAN statements can be assigned to flowchart symbols for Figure 7.1:

FORTRAN Statement Number	Flowchart Symbol Number
20, 30	1
40	2
50	3
60	4
70	5
90, 100	6
110	7

*START is not a FORTRAN STATEMENT. Instead START symbolically represents the required FORTRAN input statements, READ AND FORMAT, to enter the data.

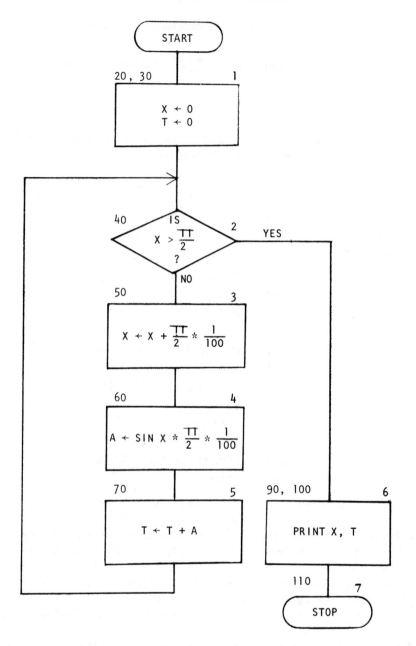

FIGURE 7.1   Area Under a Sine Wave — FORTRAN

This problem could have been originally coded in FORTRAN using a DO statement. A flowchart made from coding (DO statement) would be very similar, if not identical, to the one shown in Figure 7.1. The flowchart expresses a solution as a reference for comparison. Programming does not alter a flowchart solution.

The reader should code the problem a number of different ways and flowchart his solutions (from his coding). He should follow the procedure diagrammed.

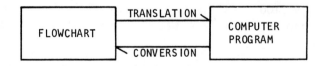

PRINTED FORM OUTPUT — COBOL

Because of its unique characteristics, the printed form output is treated in detail here using COBOL. The computer printed report is formatted with the business man in mind. Since such computer output reports are read by company executives, the report format must be clear, unambiguous, and easy to read. Specific characteristics treated in this example include the following: spacing of forms; alignment of data on forms; and overprinting of continuous strings of forms.

The example we have chosen is a computer program which was used to prepare 300 copies of the form shown in Figure 7.2. The basic form, identified as GCM–51–5/73, had been commercially printed in continuous fan-folded lots of 3200 copies with perforated tear lines between each copy. Then Computer Program TDXP01, to be described here, was used to overprint data such as: the four lines under "Ship To;" the name of the "Requisitioner," the six characters under "Shop Order Number;" and so on. The 300 copies of the form, thus modified, were for eventual use by purchasing personnel on the particular contract identified on the form under "Contract Number."

To use Computer Program TDXP01, the operator stocks his computer printer with the basic commercially preprinted forms. The program is arranged to run two copies and stop so that the operator can verify that the stock forms feed into the printer properly and that the overprinting is properly aligned in the appropriate blank spaces on the forms. If necessary, the operator can adjust the printer and ask for two more copies. Once he is satisfied, he enters an "OK" at the computer console, and the machine prints 300 copies. The flowchart for Program TDXP01 is shown in Figure 7.3.

Figures 7.4 through 7.6 show the source listing for this program. Like other programs it consists of four separate divisions. Each division is written in an English-like manner designed to decrease programming effort and to facilitate the understanding of a program by nonprogramming personnel. Each of the four divisions has a specific function:

DIVISION*	FUNCTION
1. Identification	1. Program Identification
	2. Program Documentation
2. Environment	1. Equipment Configuration
	2. Peripheral Information
3. Data	1. Input-Output Formats
	2. Intermediate work areas to convert input to output
4. Procedure	1. Contains Instructions
	2. Logic of the program within the instructions of Item 3

The program flowcharted in Figure 7.3 comprises three basic operations:

1. Display a message to the operator
2. Manually align preprint forms
3. Program execution – Print 300 forms

In the description to follow, the COBOL Procedure Division paragraphs are correlated with the flowchart symbols in Figure 7.3. The sundry details for the Identification, Environment, and Data Divisions are all necessary for program execution, yet they lack logic flow and are seldom, if ever, flowcharted. As mentioned earlier, the system analyst's package must supply complete data layout for input and output operations and is used by the programmer to fill out the Data Division paragraphs.

The program begins with an open statement and a message to the operator for his assistance in aligning the program printout with the preprint form. Next, two forms are printed and a message is displayed to the operator, Boxes 3 and 4. The ACCEPT statement (Box 5) follows, and the program execution stops. The operator reads the message, performs the required operation, and presses START; Boxes 5 and 6. If the operator is not

*See Figures 7.4 through 7.6.

RCA

This Document Number and Reference Number (if any) must appear on all Packages and Documents.

Document Number		Change No.
Reference Number		Date

Ship To

RCA
BUILDING S-805
OAKLAND ARMY BASE
OAKLAND, CALIF. 94626

Mail 3 copies of invoices to

RCA-MSR
MOORESTOWN, N.J. 08057

ATTN- ACCTS. PAYABLE
127-108

ATTN-

Requisitioner	Mail Stop Del. To	Internal Destination	Insp. Code	M or E Form	Site/Spares Number	Shop Order Number
W. SANDKUHLER		ND	0			7MM02A

Ship Via | | | | | | F.O.B.

Terms	Std. Trans. C. C.	RCA Commodity Code	The following supplements apply	These paragraphs of supplement Q apply to this order
			A	

Contract Number				DMS	Rating	Code
DAHC 60-73-C-0007				1	DXA2	

Confirmed with

If checked exempt from New Jersey Sales Tax Cert. No. 131-194-630 | Target/Estimated Price

Item	Quantity	Unit Meas.	Part Number	Rev.	Description	Unit Price	Per

CONFIRMING

PRIORITY

Delivery Schedule

Total Price.

Item	Quantity	Date Promised	Date Required	Item	Quantity	Date Promised	Date Required	Item	Quantity	Date Promised	Date Required

RCA Corporation

PURCHASING

By _____

GCM 51 5/73

FIGURE 7.2 Preprint Form for COBOL Example

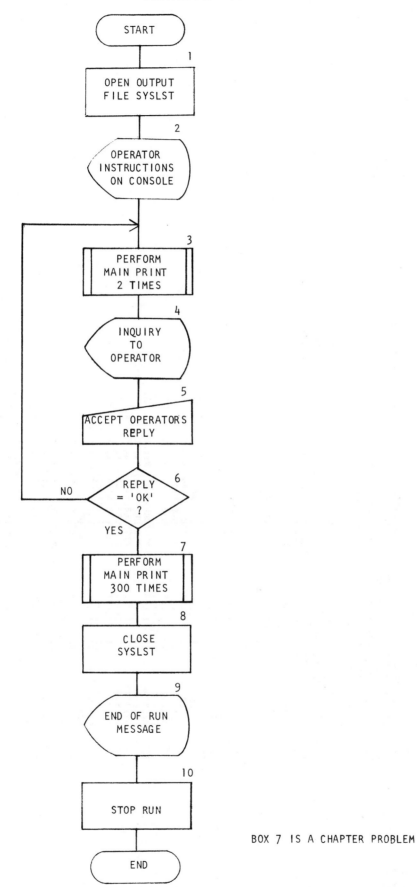

BOX 7 IS A CHAPTER PROBLEM

FIGURE 7.3 Program TDXPO1 Printed Form Output — COBOL

```
V41        COBOL COMPILATION        TDXP01         SOURCE LISTING                        06/12/74  PAGE 0001

 1   010100 IDENTIFICATION DIVISION.
 2   010200 PROGRAM-ID.     'TDXP01'.
 3   010300 AUTHOR.         DAVIDSON.
 4   010400 INSTALLATION.   C4.
 5   010500 DATE-WRITTEN.   FEBRUARY 74.
 6   010600 DATE-COMPILED.  06/27/74
 7   010700 SECURITY.       NONE.
 8   010800 REMARKS.        PRE-PRINTING OF FORM GCM 51 1½/72
 9   011000 ENVIRONMENT DIVISION.
10   011100 CONFIGURATION SECTION.
11   011200 SOURCE-COMPUTER.  UNIVAC-SERIES-70 7055.
12   011300 OBJECT-COMPUTER.  UNIVAC-SERIES-70 7055.
13   011400 INPUT-OUTPUT SECTION.
14   011500 FILE-CONTROL.
15   011600 SELECT FORM-OUT ASSIGN TO SYSLST RESERVE NO.
16   011700 DATA DIVISION.
17   011800 FILE SECTION.
18   011900 FD FORM-OUT RECORDING MODE IS U RECORD CONTAINS 86 CHARACTERS
19   012000    LABEL RECORDS ARE OMITTED
20   012100    DATA RECORD IS FORM-REC.
21   020100 01 FORM-REC.
22   020200    03 FR-CTL     PICTURE X(01).
23   020300    03 FR-DATA    PICTURE X(85).
24   020400 WORKING-STORAGE SECTION.
25   020500 01 LINE-1-10.
26   020600    03 FILLER     PICTURE X(60) VALUE SPACES.
27   020700    03 L110-DATA  PICTURE X(25).
28   020800 01 LINE-11.
29   020900    03 FILLER     PICTURE X(15) VALUE SPACES.
30   021000    03 L11-A      PICTURE X(05) VALUE 'ATTN-'.
31   021100    03 FILLER     PICTURE X(44) VALUE SPACES.
32   021200    03 L11-B      PICTURE X(07) VALUE '127-108'.
33   021300    03 FILLER     PICTURE X(14) VALUE SPACES.
34   021400 01 LINE-14.
35   021500    03 FILLER     PICTURE X(01) VALUE SPACES.
36   021600    03 L14-A      PICTURE X(13) VALUE 'W. SANDKUHLER'.
37   021700    03 FILLER     PICTURE X(18) VALUE SPACES.
38   021800    03 L14-B      PICTURE X(02) VALUE 'IND'.
39   021900    03 FILLER     PICTURE X(08) VALUE SPACES.
40   022000    03 L14-C      PICTURE X(01) VALUE '0'.
41   022100    03 FILLER     PICTURE X(27) VALUE SPACES.
42   022200    03 L14-D      PICTURE X(06) VALUE '7MH02A'.
43   022300    03 FILLER     PICTURE X(09) VALUE SPACES.
44   030100 01 LINE-18.
45   030200    03 FILLER     PICTURE X(48) VALUE SPACES.
46   030300    03 L18-A      PICTURE X(01) VALUE 'A'.
47   030400    03 FILLER     PICTURE X(36) VALUE SPACES.
48   030500 01 LINE-20.
49   030600    03 FILLER     PICTURE X(11) VALUE SPACES.
50   030700    03 L20-A      PICTURE X(17) VALUE 'DAHC 60-73-C-0007'.
51   030800    03 FILLER     PICTURE X(29) VALUE SPACES.
```

FIGURE 7.4 COBOL Listing for Preprinted Form

```
52  030900    03  L20-B      PICTURE X(09) VALUE '1' DXA2'.
53  031000    03  FILLER     PICTURE X(19) VALUE SPACES.
54  031100  01  LINE-43.
55  031200    03  FILLER     PICTURE X(02) VALUE SPACES.
56  031300    03  L43-A      PICTURE X(0R) VALUE 'PRIORITY'.
57  031400    03  FILLER     PICTURE X(14) VALUE SPACES.
58  031500    03  L43-B      PICTURE X(10) VALUE 'CONFIRMING'.
59  031600    03  FILLER     PICTURE X(51) VALUE SPACES.
60  031700  01  OPER-REPLY   PICTURE X(02).
61  040100  PROCEDURE DIVISION.
62  040200  START-1.
63  040300    OPEN OUTPUT FORM-OUT.
64  040400    DISPLAY ' PDGRAM WILL PRINT 2 FORMS' UPON CONSOLE.
65  040500    DISPLAY ' AT A TIME, ENTER NO AT  ' UPON CONSOLE.
66  040600    DISPLAY ' INQUIRY UNTIL FORMS ARE ' UPON CONSOLE.
67  040700    DISPLAY ' ALIGNED THEN ENTER OK   ' UPON CONSOLE.
68  040800  TEST-FORMS.
69  040900    PERFORM MAIN-PRINT THRU MP-EXIT 2 TIMES.
70  041000    DISPLAY 'FORMS OK?' UPON CONSOLE.
71  041100    ACCEPT OPER-REPLY FROM CONSOLE.
72  041200    IF OPER-REPLY NOT = 'OK' GO TO TEST-FORMS.
73  041300    PERFORM MAIN-PRINT THRU MP-EXIT 300 TIMES.
74  041400    CLOSE FORM-OUT.
75  041500    DISPLAY ' END OF RUN' UPON CONSOLE.
76  041600    STOP RUN.
77  060100  MAIN-PRINT.
78  060200    MOVE 'R C A' TO L110-DATA.
79  060300    MOVE LINE-1-10 TO FR-DATA.
80  060400    WRITE FORM-REC AFTER ADVANCING 0 LINES.
81  060500    MOVE 'BUILDING S-805' TO L110-DATA.
82  060600    MOVE LINE-1-10 TO FR-DATA.
83  060700    WRITE FORM-REC AFTER ADVANCING 1 LINES.
84  060800    MOVE 'OAKLAND ARMY BASE' TO L110-DATA.
85  060900    MOVE LINE-1-10 TO FR-DATA.
86  061000    WRITE FORM-REC AFTER ADVANCING 1 LINES.
87  061100    MOVE 'OAKLAND, CALIF. 94626' TO L110-DATA.
88  061200    MOVE LINE-1-10 TO FR-DATA.
89  061300    WRITE FORM-REC AFTER ADVANCING 1 LINES.
90  061400    MOVE 'RCA-MSR' TO L110-DATA.
91  061500    MOVE LINE-1-10 TO FR-DATA.
92  061600    WRITE FORM-REC AFTER ADVANCING 2 LINES.
93  061700    MOVE 'MOORESTOWN, N.J. 08057' TO L110=DATA.
94  061800    MOVE LINE-1-10 TO FR-DATA.
95  061900    WRITE FORM-REC AFTER ADVANCING 1 LINES.
96  062000    MOVE 'ATTN=ACCTS. PAYABLE' TO L110-DATA.
97  062100    MOVE LINE-1-10 TO FR-DATA.
98  062200    WRITE FORM-REC AFTER ADVANCING 3 LINES.
99  062300    MOVE LINE-11 TO FR-DATA.
100 062400    WRITE FORM-REC AFTER ADVANCING 1 LINES.
101 070100    MOVE LINE-14 TO FR-DATA.
102 070200    WRITE FORM-REC AFTER ADVANCING 3 LINES.
```

FIGURE 7.5 COBOL Listing for Preprinted Form

```
V41       COBOL COMPILATION        TDXP01              SOURCE LISTING                    06/12/74   PAGE 0003

103   070300        MOVE    LINE-18 TO FR-DATA.
104   070400        WRITE   FORM-REC AFTER ADVANCING 4 LINES.
105   070500        MOVE    LINE-20 TO FR-DATA.
106   070600        WRITE   FORM-REC AFTER ADVANCING 2 LINES.
107   070700        PERFORM LINE-SKIP 22 TIMES.
108   070800        GO TO MP-1.
109   070900   LINE-SKIP.
110   071000        MOVE SPACES TO FR-DATA.
111   071100        WRITE FORM-REC AFTER ADVANCING 1 LINES.
112   071200   MP-1.  MOVE LINE-43 TO FR-DATA.
113   071300        MOVE LINE=43 TO FR-DATA.
114   071400        WRITE FORM-REC AFTER ADVANCING 1 LINES.
115   071500   MP-EXIT.
116   071600        EXIT.
```

FIGURE 7.6 COBOL Listing for Preprinted Form

successful in his first attempt to align the preprinted forms with the computer output, he further adjusts the printer and again presses START, thus getting two more check copies. Once the operator has achieved correct form alignment he enters "OK" at the console. At this instant, the first two basic operations are completed and the third is initiated with the following statement (Box 7):

041300   PERFORM MAIN-PRINT THRU MP-EXIT 300 TIMES

A close statement (Box 8 and 041400) precedes the END OF RUN message (Box 9 and 041500) and STOP RUN (Box 10 and 041600). The remaining Procedure paragraphs, MAIN-PRINT, LINE-SKIP and MP-1, are independent routines initiated by the PERFORM statement in paragraph TEST-FORMS. The GO TO statement is used to transfer control to the paragraph name indicated and is not contained within the PERFORM statement. (See 070800.)

The spacing of forms requires special considerations for printed output. After the write instruction, the paper will advance one line; single spacing of forms will result. Additional spacing (more than one line) is accomplished by issuing an AFTER ADVANCING option with every WRITE instruction for print operations:

$$\underline{\text{WRITE}} \text{ (record-name) } \underline{\text{AFTER ADVANCING}} \begin{bmatrix} 1 \\ 2 \\ 3 \end{bmatrix} \text{LINES}$$

The notation 0 LINES is the special format used for skipping to a new page. The instruction is mechanized in the following manner. The computer must be instructed to observe page dimensions. The computer controls the sensing of various print lines by the use of a carriage control tape. This tape is attached to the printer to control the vertical format, the first and last lines of printing on page dimensions. With this tape, the computer can sense the last printed line. When the form's end is sensed, the programmer must instruct the computer to advance to the next page. The tape will sense the start of a new page as well as the end of the pervious one. Thus, through the use of the vertical carriage format control tape, the programmer can test for the end of a form and then skip to a new page when necessary.

QUADRATIC – ALGOL

The quadratic equation is a classic classroom lecture and text book problem. The equation solution encompasses completing the square on one side of an equal sign and taking the square root of each equation on both sides of the equal sign.

$$ax^2 + bx + c = 0 \qquad\qquad \text{Quadratic}$$

$$x^2 + \frac{b}{a}x = -\frac{c}{a}$$

$$x^2 + \frac{b}{a}x + \left(\frac{b}{2a}\right)^2 = \frac{b^2}{4a^2} - \frac{c}{a} \qquad\qquad \text{Completing the square}$$

$$\left(x + \frac{b}{2a}\right)^2 = \frac{b^2 - 4ac}{4a^2}$$

$$x + \frac{b}{2a} = \pm\frac{\sqrt{b^2 - 4ac}}{2a} \qquad\qquad \text{Taking the square root}$$

$$x = \frac{-b \pm \sqrt{b^2 - 4ac}}{2a} \qquad\qquad \text{Quadratic solution}$$

The solution of the quadratic holds for all values of a, b, and c, except for a = 0.
We now consider a program to solve the quadratic equation:

$$ax^2 + bx + c = 0$$

with arbitrary real coefficients a, b, and c. The ALGOL program serves as an example of the self-embedding capability of the high level language. The program was run on the Burroughs B6700 and the printouts include a time shared terminal printout without indentations (Figure 7.7) and a computer page printout (120 columns – Figure 7.8) with indentation. The block structured feature of the ALGOL programming language is shown in the computer page printout. The Burroughs B6700 is structured to reflect the ALGOL like block structure.[1] The problem example features the following items:

1. Flowcharting a high level programming language listing.
2. Push Down Stack – both software and hardware.
3. Introduction to ALGOL and an algorithmic structured computer, Burroughs B6700.

The quadratic example is detailed using the remote terminal printout (Figure 7.7) and the flowchart in Figure 7.9. The terminal printout has very little indentation and the reader is shown how the nested or self embedding ALGOL properties are presented in flowcharts. The printout listing of comment statements, parametric declarations, file assignments, and input/output formats up to Line Number 920 are all summarized using the start symbol. The A, B, and C values (Box 1) represent Line 1100. The assignment Line 1200 is shown as Box 3. The test $A \neq 0$ is shown as Box 2 and Line 1300. If the test for $A \neq 0$ is true we begin and test for the $DISC \neq 0$; Box 4 and Line 1500. In order to make the test, the DISC must be computed and evaluated. The computation is performed in Box 3. Box 3 could not be left out in the flowchart. The value for the DISC was declared prior to its use in the listing. The value for the DISC could be computed at any time provided A, B, and C are available; after readin. The most logical location for Box 3 is the one shown in the flowchart although it could be located right after readin; Box 1. Next, a test is performed to see if $DISC \geqslant 0$ (Box 4 and Line 1500). If true R is used to represent the square root of DISC; Box 5 and Line 1700. A test is made on the B value; Box 6 and Line 1800. If true there are two real roots:

$$\text{Line 2000 X1} \; := \; -(B + \text{SIGN}(B) * R)/(2*A) \; ;$$

$$\text{Line 2100 X2} \; := \; (C/(A*X1)$$

At this point, a solution is available for the logic path flow for Boxes 1 thru 7, as noted in Line 2200. At Line 2300, we go backward to the first encountered uncompleted decision symbol, Box 6, and flowchart Line 2500. The ELSE statement in Line 2300 is the path taken for $B = 0$ in Box 6. Box 8 represents Lines 2500 and 2600. Boxes 7 and 8 are the real part solutions (Line 2800) for this quadratic. At Line 2900, we go backward to the first encountered uncompleted decision symbol; in this case it would be Box 4. Next, Lines 3100, 3200, and 3300 are flowcharted and represented as Boxes 12, 13 and 14 respectively. At Line 3400, a solution is available for conjugate complex roots for the logic path flow for Boxes 1 thru 4 and Boxes 12 thru 14. At Line 3600, we go backward to the first encountered uncompleted decision symbol. In this case it would be Box 2. Manually, the listing has been flowcharted to a depth of 3 (a stack of 3) and the process has been reversed (a stack of 3 has been unloaded or emptied). At Box 9 and Line 3700, the process is repeated. The student can check the listing; Lines 3800 thru 4600 to conform with Boxes 9 through 11. The printout operation, Box 15 and Lines 4800 and 4900, are initiated when the particular END statement is processed.

Throughout this book, the start message symbol is used to introduce the logic flow. The quadratic output listing is used to exemplify this operation. The work flow statements and the job control language representing the equivalent start procedure are given in Figure 7.8, Parts 1 through 3. Figure 7.8 is annotated to show how the quadratic example was run. The quadratic program is initially compiled and the six sets of inputs are run as six tasks sequentially. The job summary lists (Part 2) 48 cards read and 64 lines printed for the quadratic, and one card read and one line printed for each quadratic run. The statistics for the work flow are given in Figure 7.8, Part 3. The equivalent start operation in Figure 7.9 would be Parts 1 through 3 of Figure 7.8 excluding the program statements.

The quadratic program in ALGOL is given in Figure 7.8, Part 4. The figure is annotated for illustration purposes. In practice, the procedure is to extensively comment the processing statements for explanation, documentation, and program maintenance. The ALGOL nesting configuration is not difficult to see with the indentation of a computer output listing (Figure 7.8, Part 3).

```
LIST
100 BEGIN COMMENT : THE SOLUTION OF THE QUADRATIC
200 EQUATION A*X*X+B*X+C=0 WITH ARBITRARY REAL
300 COFFFICIENTS A,B,C;
400 REAL A,B,C,X1,X2,REALPART,IMAGINARYPART,
500 CONTRADICTION,DISCR,R;
600 FILE PRINT(KIND=REMOTE);
700 FILE CARD(KIND=READER,MAXRECSIZE=14);'
800 FORMAT INPUT(3F10.3);
900  FORMAT OUTPUT("X1=",F10.4," X2=",F10.4,
901  /,
910  "REAL PART=",F10.4," IMAGINARY PART=",F10.4,
920  /,"CONTRADICTION=",F10.4);
1100   A:=2;B:=17;C:=23;
1200 DISCR:=B*B-4*A*C;
1300 IF A NEQ 0 THEN
1400 BEGIN
1500 IF DISCR GEQ 0 THEN
1600 BEGIN
1700 R:=SQRT(DISCR);
1800  IF B NEQ 0 THEN
1900 BEGIN
2000 X1:=-(B+SIGN(B)*R)/(2*A);
2100 X2:=C/(A*X1);
2200 END B
2300 ELSE
2400 BEGIN
2500 X1:=R/(2*A);
2600 X2:=-X1;
2700 END
2800 END TWO REAL SOLUTIONS
2900 ELSE
3000 BEGIN
3100 R:=SQRT(ABS(DISCR));
3200 REALPART:=-B/(2*A);
3300 IMAGINARYPART:=R/(2*A);
3400 END TWO CONJUGATE COMPLEX SOLUTIONS
3500 END A NOT EQUAL ZERO
3600  ELSE
3700 BEGIN
3800  IF B NEQ 0 THEN
3900 BEGIN
4000 X2:=-C/B;
4100 END ONE REAL SOLUTION
4200 ELSE
4300 BEGIN
4400 CONTRADICTION:=C;
4500 END SOLUTION UNDETERMINED
4600 END A EQUAL TO ZERO
4700 ;
4800  WRITE(PRINT,OUTPUT,X1,X2,REALPART,IMAGINARYPART,
4810  CONTRADICTION);
4900 END.
5000
#
1100 A:=1;B:=.5;C:=10;
SAVE
#UPDATING
#WORKFILE SAVED; OLD FILE REMOVED
COMPILE Q ALGOL
#COMPILING 2258
#ET=10.0 PT=1.2 IO=0.7
RUN Q
#RUNNING 2261
X1=    0.0000 X2=    0.0000
REAL PART=   -0.2500 IMAGINARY PART=    3.1524
CONTRADICTION=    0.0000
#ET=14.5 PT=0.2 IO=0.2


BYE
#END SESSION 2066  ET=36:26.3 PT=9.6 IO=6.3
#USER = MSDD  16:12:03  06/06/74
```

FIGURE 7.7 ALGOL Listing for Quadratic Example

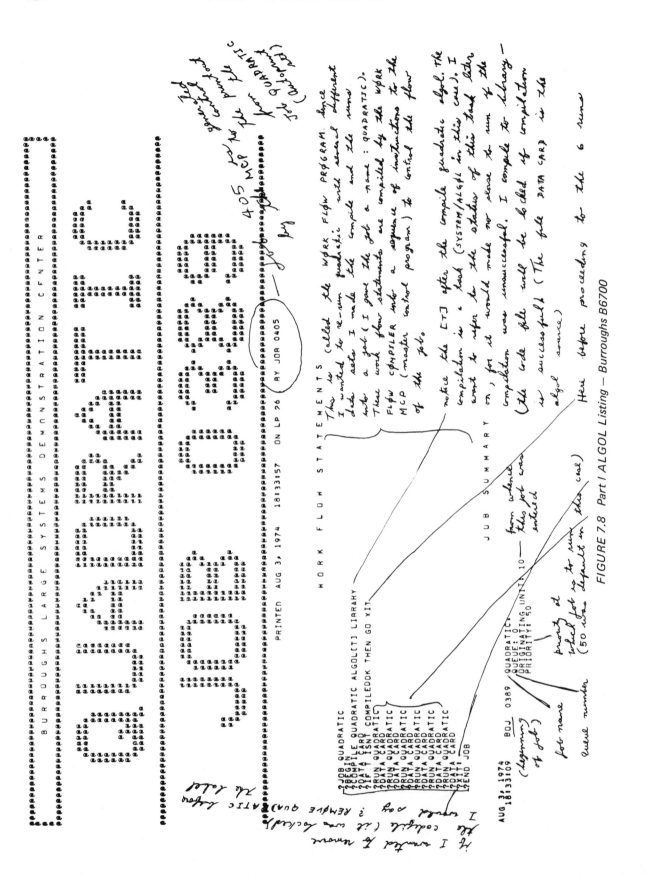

FIGURE 7.8 Part I ALGOL Listing — Burroughs B6700

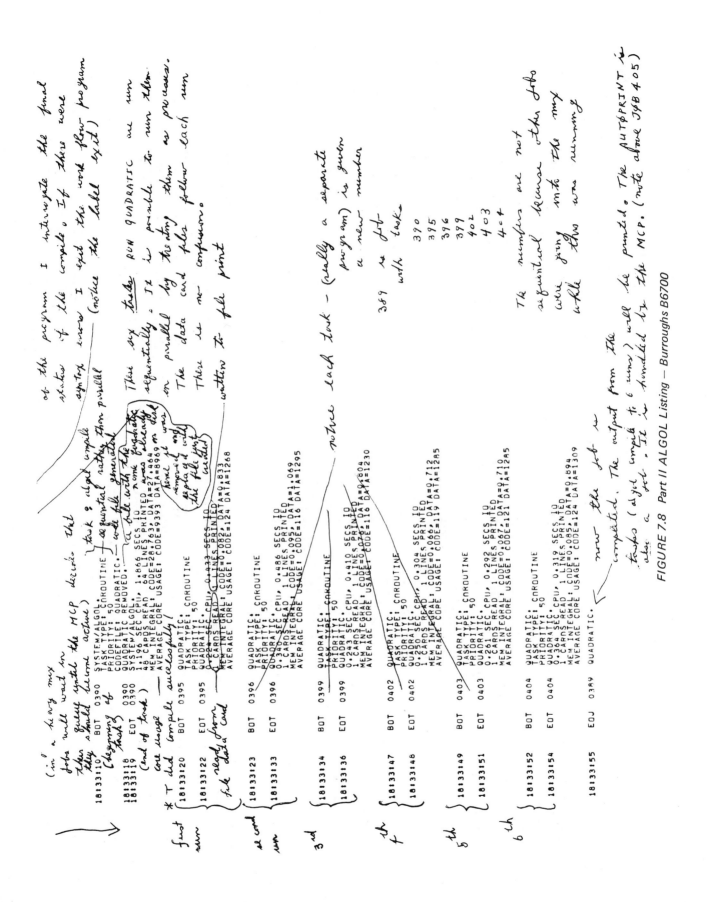

FIGURE 7.8 Part II ALGOL Listing — Burroughs B6700

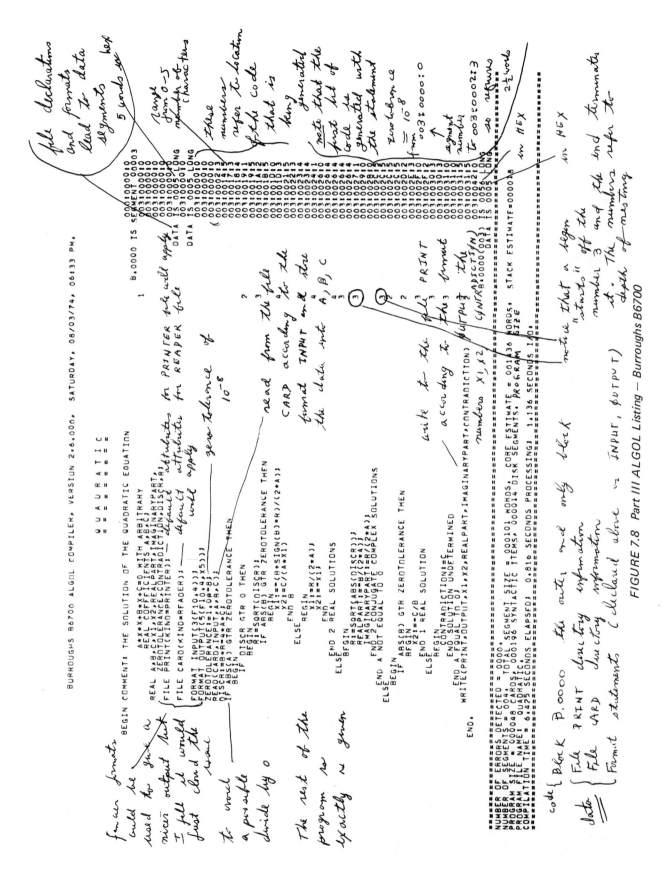

FIGURE 7.8 Part III ALGOL Listing — Burroughs B6700

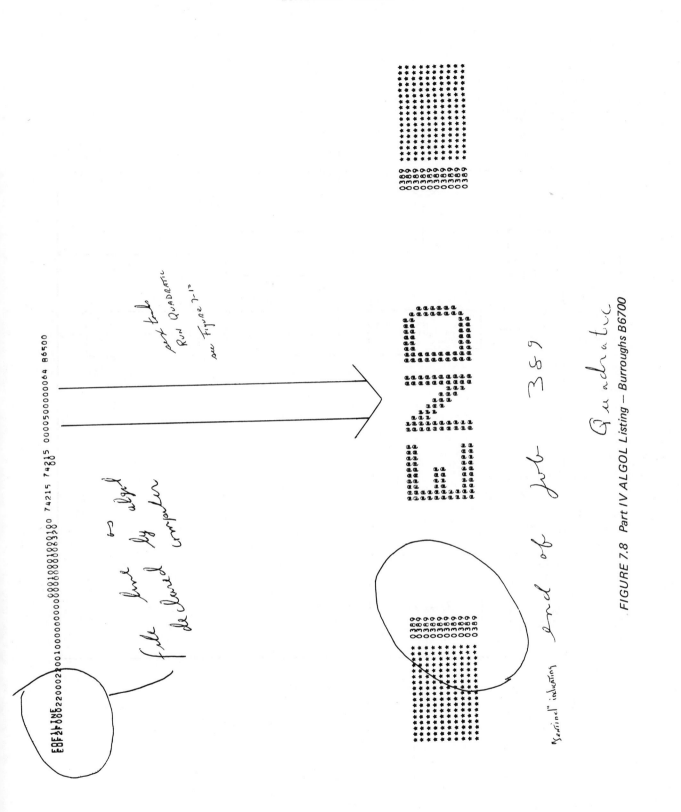

FIGURE 7.8 Part IV ALGOL Listing – Burroughs B6700

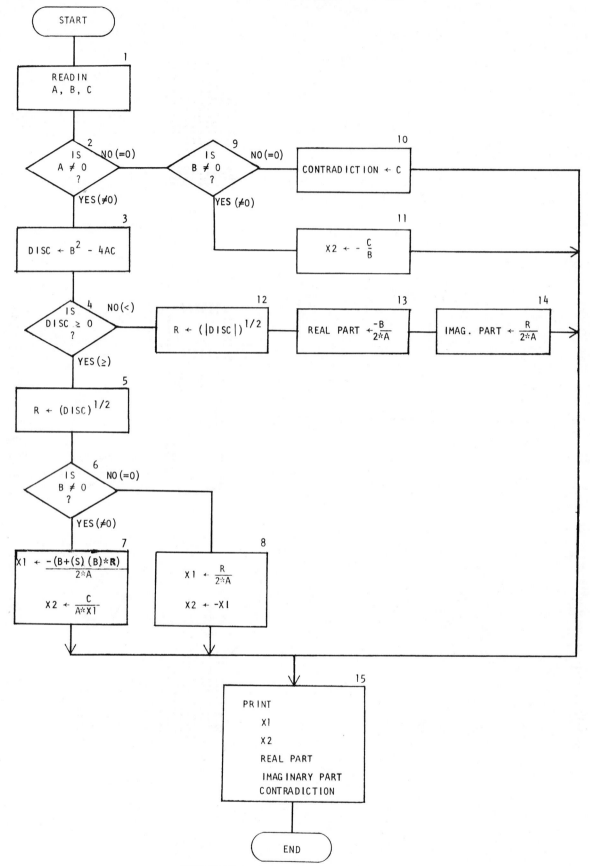

FIGURE 7.9   Quadratic – ALGOL

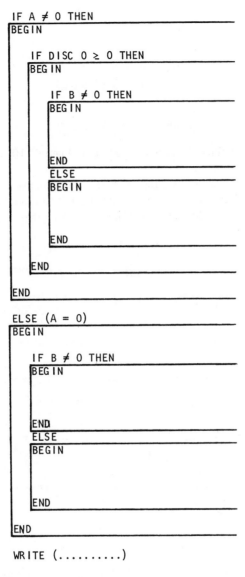

```
BEGIN

    IF A ≠ 0 THEN
    BEGIN

        IF DISC 0 ≥ 0 THEN
        BEGIN

            IF B ≠ 0 THEN
            BEGIN



            END
            ELSE
            BEGIN



            END

        END

    END

    ELSE (A = 0)
    BEGIN

        IF B ≠ 0 THEN
        BEGIN



        END
        ELSE
        BEGIN



        END

    END

    WRITE (.........)

END
```

In the outline form shown, the ALGOL block structuring is easily identifiable. The first block structure is formed with the BEGIN that is associated with the declaration of a list of parameters. Included in the statement sequence is the program title, input/output devices, and input/output formats. The printout statement, WRITE, is located in the first block structure. The degree of nesting is given in the output listing, Part 3 of Figure 7.8, as a separate (single digit) column just right of the program statements and left of the card numbering and program identification. On the same sheet the program size is given and the figure correlates with the compilation process of the job summary in Part 2.

The ALGOL program (Figure 7.8, Parts 1 thru 4) was run with six sets of input data

	A	B	C
1st	1	2	3
2nd	1	0	–1
3rd	0	0	1
4th	–1	2.5	3.9
5th	4	4	1
6th	1.	2.5	3.9

and their respective outputs are assembled without subheadings in Figure 7.10. The WRITE statement must be consulted for parametric identification.

The quadratic runs are listed as six tasks in Figure 7.8, Part 2 and were executed after compilation as a six set sequence in Figure 7.8, Part 4, with an END OF JOB sentinel. Each data run resulted in a single line printout of computed results sandwiched between a file print as declared in the program task Quadratic and an END OF PRINT

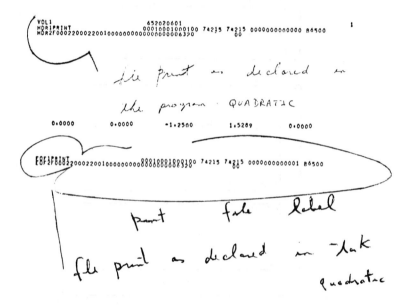

The reader is introduced to flowcharting from a program listing. This is a bottom up operation. This flowcharting process is sometimes called documentation, and there are several commercial proprietary software packages available (e.g., AUTOFLOW*). Figure 7.7 is a typical example of interactive remote terminal output. Unless the Computer Data Center buffer output format is compatible with a remote teletype terminal, the normal 120 column printout will repeatedly hit the last symbol location of line until a carriage return and advance a line signal(s) are issued. Therefore, most teletype print outputs have a common or single margin on the left side to obtain the maximum symbol capacity per line.

PRIME NUMBER SIEVE – PL/1

Prime numbers are presented as a pattern of asterisk symbols. Prime numbers usually scan to appear "at random" among the integers; this program demonstrates that this is not quite so. In this example, the program flowchart is presented in two forms:

1. ANSI Flowchart Symbols
2. Structured Flowchart Symbols[2]

*Copyright by Applied Data Research, Inc., Princeton, New Jersey

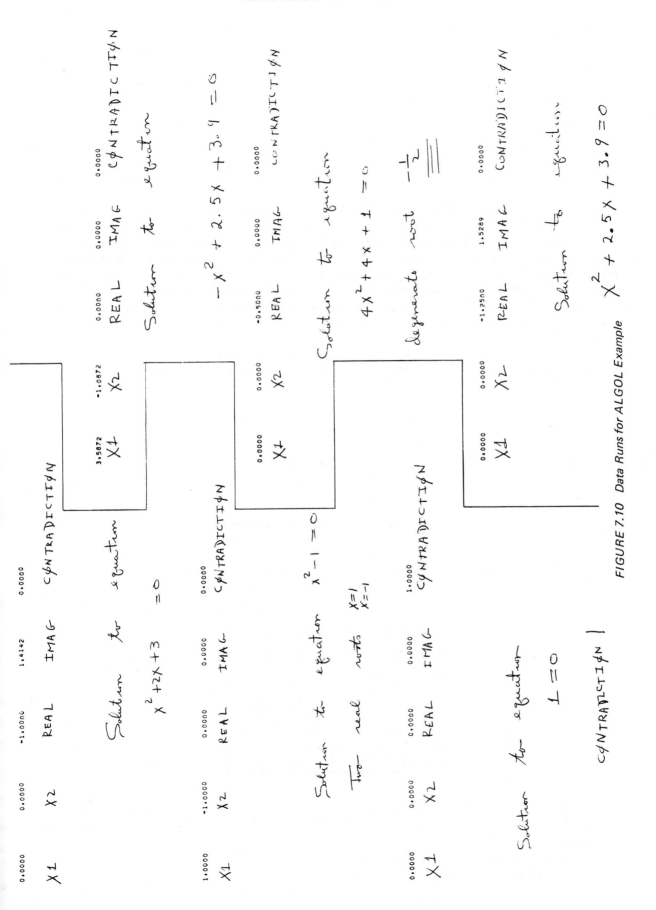

FIGURE 7.10 Data Runs for ALGOL Example

The program flowchart was drawn in the structured flowchart format in Figure 7.11. As shown here, the program consists of four major operations:

1. SIEVE
2. INDENT
3. TWIST
4. PRINT

with an initialization operation as part of TWIST.

The basic control operations such as process, iteration, and decision functions are compared with another set of structured flowchart symbols, and the start and end ANSI symbols are represented in a START and END block in Figure 7.12.

The Figure 7.11 flowchart is quite similar to the ALGOL structure presented earlier, and the nesting structure of both are alike. The Figure 7.11 flowchart is redrawn using ANSI symbols in Figure 7.13. The major operation labels (SIEVE, IDENT, TWIST, and PRINT) are retained, and the nested loop configuration is visible to the reader.

The Prime Number Sieve program was run on an IBM 360/65 system using the DATA 100 CORP REMOTE BATCH System. The job control language and statistics are given in Figure 7.14, the PL/1 listing is given in Figure 7.15, and the prime number "radar tracks" are shown in Figure 7.16 (computer printout).

The program listing is relatively short and the program details are referenced to the listing. Reference to the Figure 7.13 flowchart is left to the reader as an exercise.* The vector SIV is initially set to all asterisks ("*") (Line 4), then the usual prime sieve (Lines 5-9) sets the elements of composite subscript of SIV to blanks. Positions 0 through 9 are set to the characters "0" to "9", respectively, so we can more clearly see what is going on (Lines 10 through 12). The next action (Lines 16 through 28) wraps the vector SIV into a matrix (two dimensional array) PAT as shown:

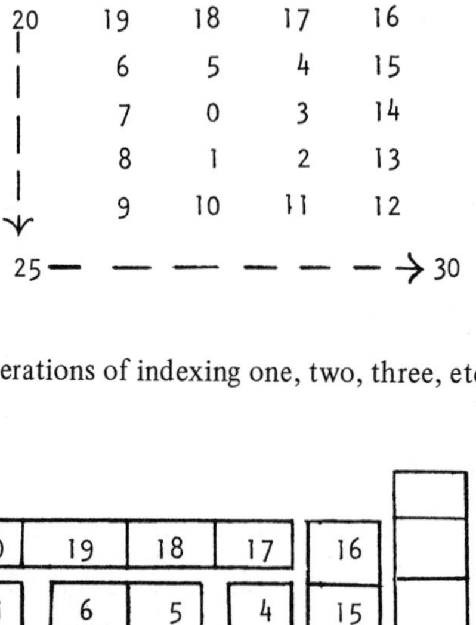

```
20    19    18    17    16
 |     6     5     4    15
 |     7     0     3    14
 |     8     1     2    13
 |     9    10    11    12
 ↓
25 —  —  —  —  —  → 30
```

by DO Loop (Line 18 and Line 23) operations of indexing one, two, three, etc., starting with the symbol of zero "0".

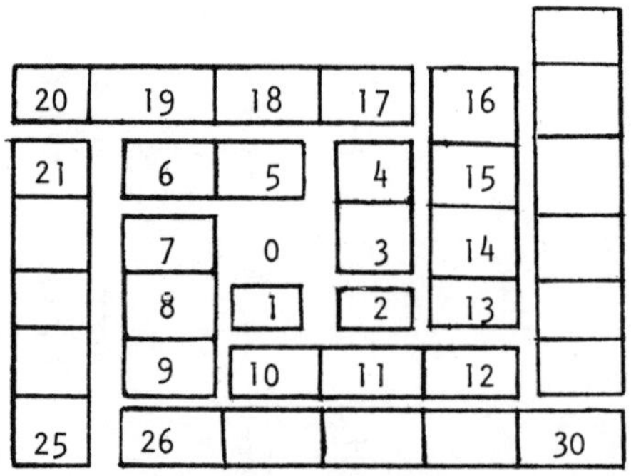

20	19	18	17	16	
21	6	5	4	15	
	7	0	3	14	
	8	1	2	13	
	9	10	11	12	
25	26				30

The vector SIV is converted to a matrix notation after the completion of each DO loop operation within TWIST; Line 21 for DO Line 18 and Line 26 for DO Line 23. PAT is printed (Line 29) showing the "radar tracks" of the prime numbers in Figure 7.16.

*Why are the decision symbols located at they are in Figure 7.13?

(Hint – See PL/1 Language Characteristics)

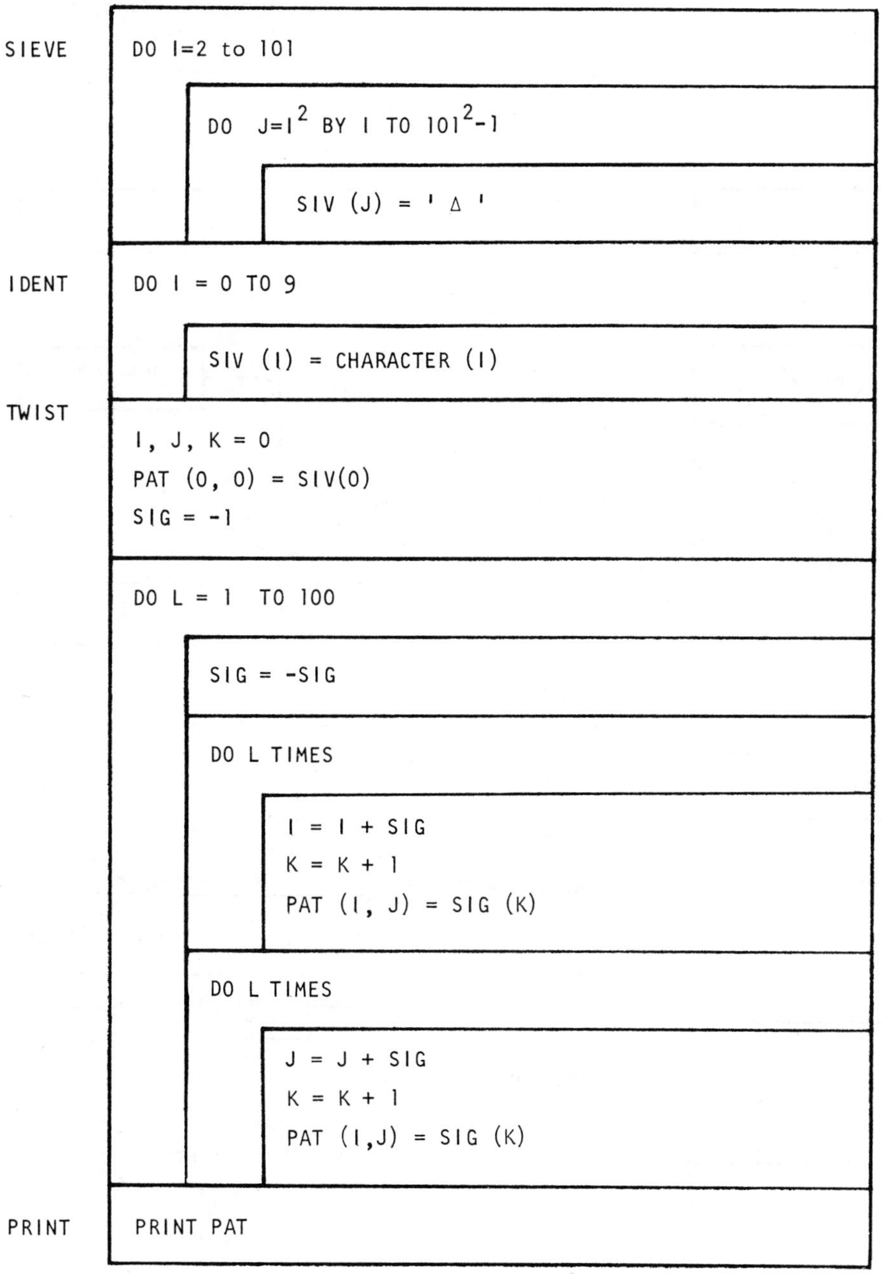

SIEVE	DO I=2 to 101

$$\text{DO} \quad J=I^2 \text{ BY I TO } 101^2-1$$

SIV (J) = ' Δ '

IDENT	DO I = 0 TO 9

SIV (I) = CHARACTER (I)

TWIST	I, J, K = 0

PAT (0, 0) = SIV(0)
SIG = -1

DO L = 1 TO 100

SIG = -SIG

DO L TIMES

I = I + SIG
K = K + 1
PAT (I, J) = SIG (K)

DO L TIMES

J = J + SIG
K = K + 1
PAT (I,J) = SIG (K)

PRINT	PRINT PAT

FIGURE 7.11   Prime Number Sieve — PL/1

STRUCTURE   STRUCTURE
FLOWCHART   FLOWCHART
SYMBOLS   CONTROL OPERATIONS
(REFERENCE 2)   (ANSI SYMBOLS)
                                             (REFERENCE 3)

BASIC PROGRAMMING OPERATIONS

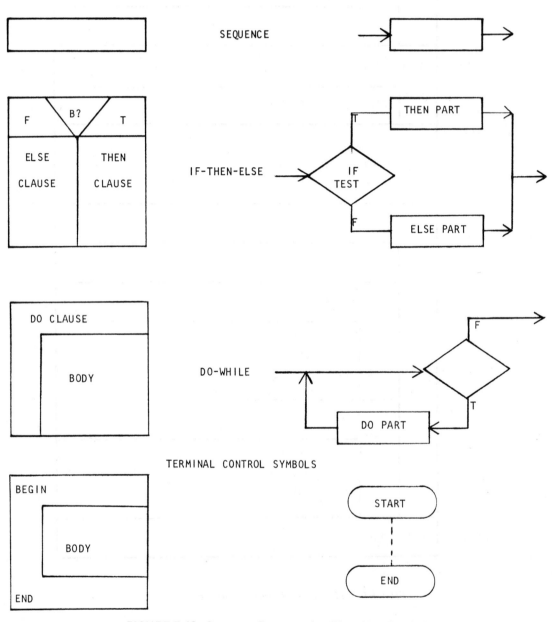

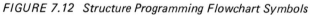

FIGURE 7.12 Structure Programming Flowchart Symbols

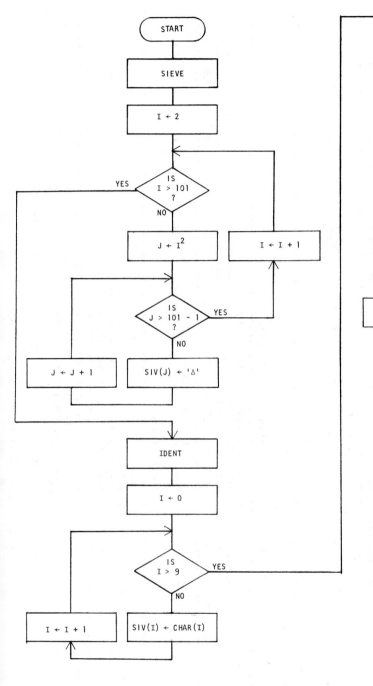
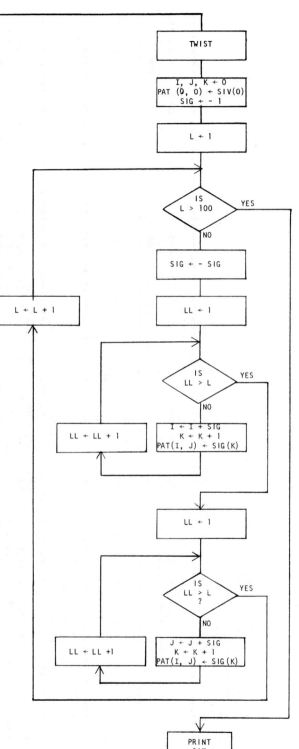

FIGURE 7.13 *Prime Number Sieve — PL/1*

```
//T35104PG JOB (W2A04),'600,ANDERSON',CLASS=A,MSGLEVEL=(0,0),
//  TIME=(0000,10) DEFAULT TIME SET BY HASP                          JOB 845
IEF142I - STEP WAS EXECUTED - COND CODE 0000
IEF373I STEP /PLI   / START 74210.1615
IEF374I STEP /PLI   / STOP  74210.1615 CPU    0MIN 01.51SEC MAIN 100K LCS    0K
**********************************  PACES DATA ACQUISITION SYSTEM  **********************************
* STEP NAME   PLI        START TIME 16.15.18.00    MAIN CORE REQD    100 K    LCS CORE REQD    0 K    STEP CPU    00.00.01.51 *
* PGM NAME    IELOAA     STOP  TIME 16.15.54.55    MAIN CORE USED    100 K    LCS CORE USED    0 K    JOB  CPU    00.00.01.51 *
* DISPATCH PRTY 123      ELAP. TIME 00.00.36.55    MAIN CORE BORRWD    0 K    LCS CORE BORRWD  0 K    CONDITION CODE    0000  *
**********************************  EXCP STATISTICS  **********************************
* UNIT   EXCP COUNT    UNIT   EXCP COUNT    UNIT   EXCP COUNT    UNIT   EXCP COUNT    UNIT   EXCP COUNT *
* D33       167        2C0        51        D0F       266        46                                    *
                                                         EXCP TOTAL    2
IEF142I - STEP WAS EXECUTED - COND CODE 0000
IEF373I STEP /LKED  / START 74210.1615
IEF374I STEP /LKED  / STOP  74210.1616 CPU    0MIN 00.56SEC MAIN  98K LCS    0K
**********************************  PACES DATA ACQUISITION SYSTEM  **********************************
* STEP NAME   LKED       START TIME 16.15.54.84    MAIN CORE REQD    100 K    LCS CORE REQD    0 K    STEP CPU    00.00.00.56 *
* PGM NAME    IEWL       STOP  TIME 16.16.14.84    MAIN CORE USED     98 K    LCS CORE USED    0 K    JOB  CPU    00.00.02.07 *
* DISPATCH PRTY 123      ELAP. TIME 00.00.20.00    MAIN CORE BORRWD    0 K    LCS CORE BORRWD  0 K    CONDITION CODE    0000  *
**********************************  EXCP STATISTICS  **********************************
* UNIT   EXCP COUNT    UNIT   EXCP COUNT    UNIT   EXCP COUNT    UNIT   EXCP COUNT    UNIT   EXCP COUNT *
* 167        60        2C0         0        17         17        D33       42        2C0       52        *
                                                         EXCP TOTAL    171
IEF142I - STEP WAS EXECUTED - COND CODE 0000
IEF373I STEP /GO    / START 74210.1616
IEF374I STEP /GO    / STOP  74210.1616 CPU    0MIN 03.61SEC MAIN  54K LCS    0K
**********************************  PACES DATA ACQUISITION SYSTEM  **********************************
* STEP NAME   GO         START TIME 16.16.15.04    MAIN CORE REQD    100 K    LCS CORE REQD    0 K    STEP CPU    00.00.03.61 *
* PGM NAME    PGM=*.DD   STOP  TIME 16.16.37.38    MAIN CORE USED     54 K    LCS CORE USED    0 K    JOB  CPU    00.00.05.68 *
* DISPATCH PRTY 123      ELAP. TIME 00.00.22.34    MAIN CORE BORRWD    0 K    LCS CORE BORRWD  0 K    CONDITION CODE    0000  *
**********************************  EXCP STATISTICS  **********************************
* UNIT   EXCP COUNT    UNIT   EXCP COUNT    UNIT   EXCP COUNT    UNIT   EXCP COUNT    UNIT   EXCP COUNT *
* 2C0         0        D33       101                                                                   *
                                                         EXCP TOTAL    101
IEF375I JOB /T35104PG/ START 74210.1615
IEF376I JOB /T35104PG/ STOP  74210.1616 CPU    0MIN 05.68SEC
**********************************  PACES DATA ACQUISITION SYSTEM  **********************************
* JOB LOG NUMBER - T35104PG 74210 16.14.57.02       CPU TIME   00.00.05.68    INITIATION TIME    16.15.18.00 *
* PROGRAMMER  600,ANDERSON                           INIT DATE  07/29/74  74.210   TERMINATION TIME   16.16.41.03 *
* ACCTG DATA  W2A04                                  TERM DATE  07/29/74  74.210   ELAPSED TIME       00.01.23.03 *
* OS-MVT REL 21.7                                    PRIORITY   07            COMPLETION STATUS    C0000 *
* SYSTEM ID  17 - EE                                 CLASS      A                                       *
```

FIGURE 7.14 Job Control Language — Sieve PL/1

```
PL/I OPTIMIZING COMPILER          WIRL:    PROCEDURE OPTIONS(MAIN);


                        SOURCE LISTING


    STMT LEV NT


      1       0   WIRL:    PROCEDURE OPTIONS(MAIN);

                           /*    THE PRIME NUMBERS WITH A TWIST.....    */
                           /*    CODED JULY 1974 BY PETER G. ANDERSON   */

      2    1  0            DECLARE(I, J, K, L, LL, SIG) FIXED BINARY;
      3    1  0            DECLARE PAT(-50:50,-50:50) CHAR(1);
      4    1  0            DECLARE SIV (0:101*101-1) CHAR(1) INIT((10201)(1)'*');

      5    1  0   SIEVE OF ERATOSTHENES:
                           DO I = 2 TO 101;
      6    1  1                DO J = I*I BY I TO 101*101-1;
      7    1  2                    SIV(J) = ' ';
      8    1  2                END;
      9    1  1            END;

     10    1  0   IDENTIFY_THE_FIRST_TEN_NUMBERS:
                           DO I = 0 TO 9;
     11    1  1                SIV(I) = SUBSTR('0123456789', I+1, 1);
     12    1  1            END;

     13    1  0   TWIST THEM UP:
                           I, J, K = 0;
     14    1  0            PAT(0,0) = SIV(0);
     15    1  0            SIG = -1;

     16    1  0            DO L = 1 TO 100;
     17    1  1                SIG = -SIG;
     18    1  1                DO LL = 1 TO L;
     19    1  2                    I = I + SIG; K = K + 1;
     21    1  2                    PAT(I,J) = SIV(K);
     22    1  2                END;

     23    1  1                DO LL = 1 TO L;
     24    1  2                    J = J + SIG; K = K + 1;
     26    1  2                    PAT(I,J) = SIV(K);
     27    1  2                END;
     28    1  1            END;

     29    1  0   PRINT THE_TWISTED_PICTURE:
                           PUT EDIT ( PAT ) ( COL(15), 101 A );

     30    1  0   END      WIRL;
```

FIGURE 7.15 PL/1 Listing for Prime Number Sieve

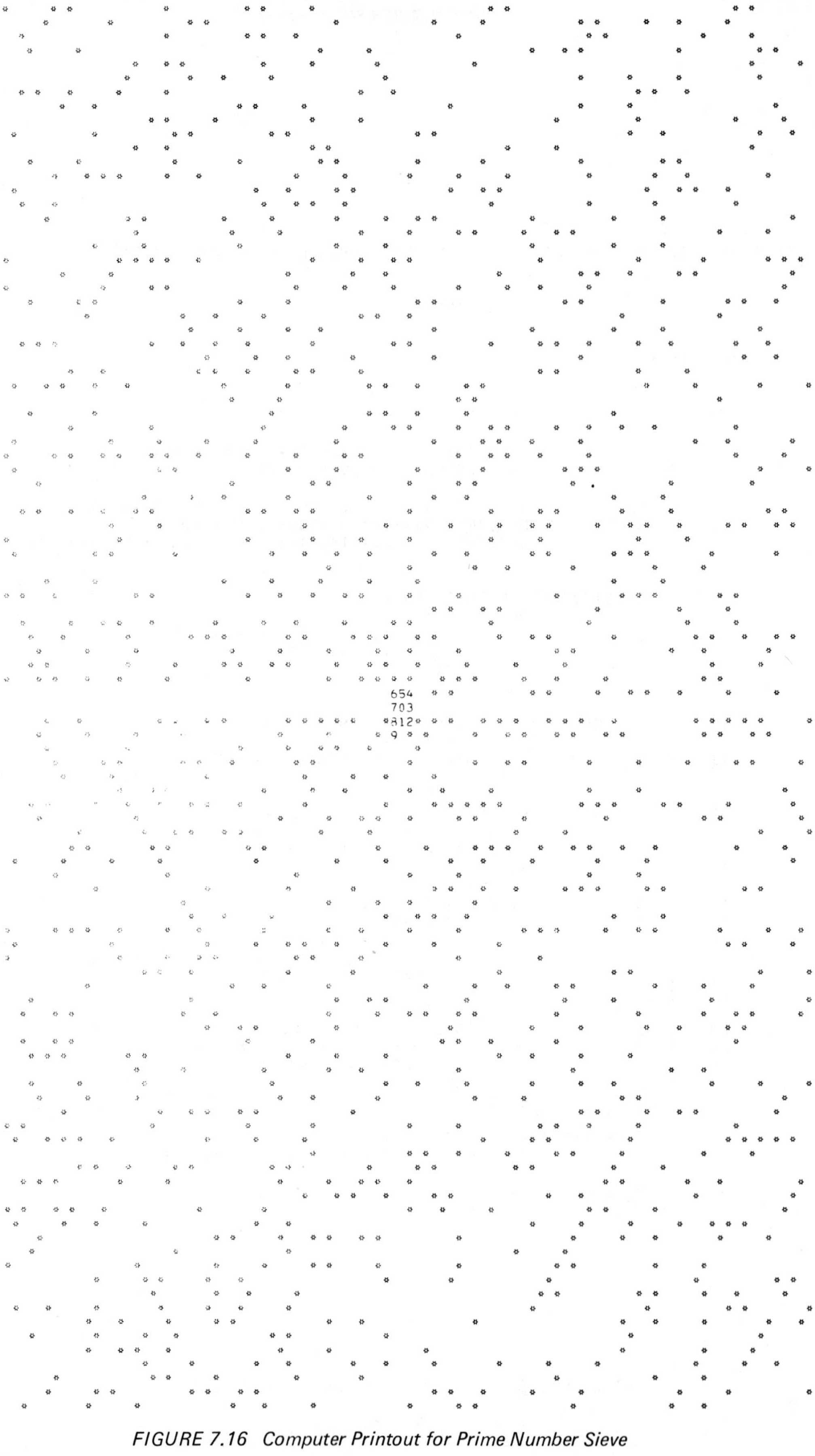

FIGURE 7.16   Computer Printout for Prime Number Sieve

LARGER OF TWO — FOCAL

Minicomputers and microprocessors are receiving considerable attention at the moment and the next two examples are presented with this in mind.

The first example is a program written in FOCAL.* It accepts entries only valid for numbers between zero and seven inclusive. If a number is greater than seven ($X > 7$) or negative ($X < 0$), another value must be read in. This process continues until the number is acceptable ($0 \leqslant X \geqslant 7$).

The program flowchart and the FOCAL listing are both given in Figure 7.17. Here the statement numbers are assigned to the flowchart symbols. The FOCAL language accepts abbreviations for reserved words or symbol strings, (e.g., G for GOTO and S for SET). The instruction "DO7" directs the program to execute all instructions beginning with a 7 in the line number. For clarity, this could be rephrased "DO 7.1" which would produce the same effect in a run (of this particular program).

The convenience of any high level language is never fully appreciated until a program written in the high level language can be compared with the corresponding object code listing. The assembly level program for the described FOCAL program is shown in Figure 7.18 and its corresponding flowchart in Figure 7.19. The comparison between writing the program in assembly language and in FOCAL is quite clear. The assembly program performs once from a starting address 0200_8 - - if the program is to be re-executed, the initial value of SHAZAM must be reset to - 4 (set 7774_8 in location 524); then load address 0200 and start.

In the assembly program flowchart (Figure 7.19) the series of flowchart symbols represent inline coding. The loop operation is not visible. At this level of flowchart representation, the general case uses two of the three structured program elements. The loop configuration is not present.

The acceptable entries were limited to numbers between zero and seven in order to keep this example from becoming overly complex.

(Text continues on page 212.)

---

*FOCAL (Formulating On Line Calculations in Algebraic Language) is a trademark of Digital Equipment Corporation. The FOCAL system is used on PDP-8 series computers.

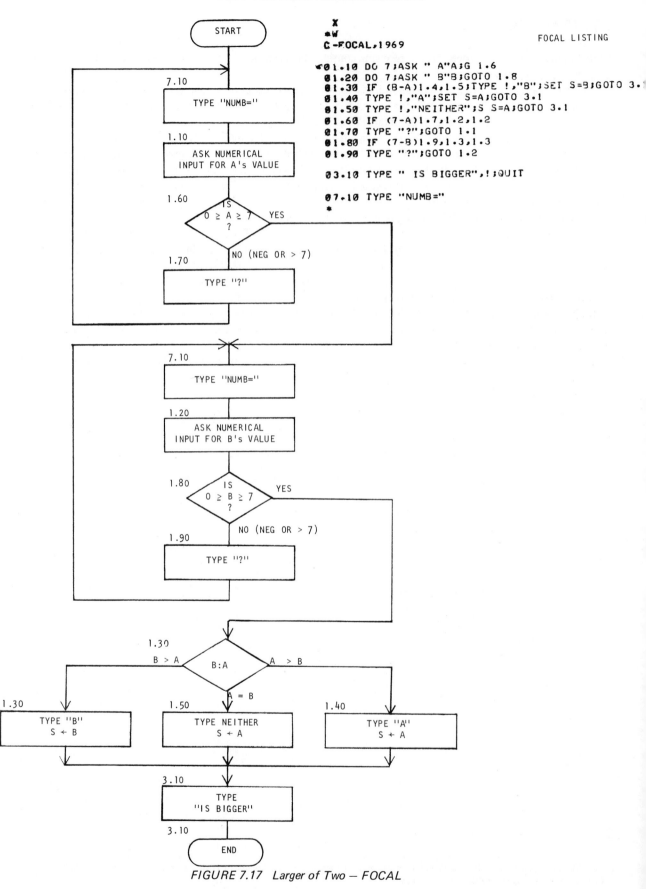

```
    X
   *W
C -FOCAL,1969                              FOCAL LISTING

01.10 DO 7;ASK " A"A;G 1.6
01.20 DO 7;ASK " B"B;GOTO 1.8
01.30 IF (B-A)1.4,1.5;TYPE !,"B";SET S=B;GOTO 3.
01.40 TYPE !,"A";SET S=A;GOTO 3.1
01.50 TYPE !,"NEITHER";S S=A;GOTO 3.1
01.60 IF (7-A)1.7,1.2,1.2
01.70 TYPE "?";GOTO 1.1
01.80 IF (7-B)1.9,1.3,1.3
01.90 TYPE "?";GOTO 1.2

03.10 TYPE " IS BIGGER",!;QUIT

07.10 TYPE "NUMB="
   *
```

FIGURE 7.17 Larger of Two — FOCAL

```
        *200
START,  CLA CLL       /CLEAR ACCUMULATOR & LINK
        JMS QUEST     /GO TO QUESTION SUBROUTINE
        CMS A         /LET'S GET A'S VALUE
        CLA CLL       /CLEAR ACCUMULATOR AND LINK
        TAD TEMPY     /GET THE ENTRY
        DCA NUMA      /ASSIGN ITS VALUE TO A
        JMS CRLF      /DO A CARRIAGE RETURN AND LINE FEED
        CLL           /CLEAR LINK
        JMS QUEST     /QUESTION SUBROUTINE AGAIN
        JMS B         /NOW FOR B'S VALUE
        CLA CLL       /CLEAR LINK AND ACCUMULATOR
        TAD TEMPY     /GET THE NEXT ENTRY
        DCA NUMB      /ASSIGN ITS VALUE TO B
        JMS CRLF      /EXECUTE A CARRIAGE RETURN-LINE FEED
        CLA CLL       /CLEAR LINK AND ACCUMULATOR
        TAD NUMB      /TAKE B
        CIA           /FROM
        TAD NUMA      /A
        SNA           /UNLESS BOTH ARE EQUAL
        JMP EQ        /IN WHICH CASE WE ACKNOWLEDGE SAME
        SMA           /CHECK TO SEE WHETHER A OR B IS BIGGER
        JMP AQ        /IF A, ACKNOWLEDGE IT;
        JMP BQ        /OTHERWISE, ACKNOWLEDGE B
NUMA,   0             /A
NUMB,   0             /B

QUEST,  0             /QUESTION SUBROUTINE
        CLA CLL       /CLEAR LINK AND ACCUMULATOR
        TLS           /HOIST PRINTER FLAG
        TAD CHARAC    /CREATE AUTOINDEX REGISTER
        DCA IRI       /FOR CHARACTERS
        TAD I IRI     /SET UP COUNTER FOR
        DCA COUNT     /COUNTING CHARACTERS
        JMS TYPE      /GET A CHARACTER
        ISZ COUNT     /TYPE IT
        JMP THEN      /DONE?
        JMP I QUEST   /NO: GET ANOTHER
                      /YES: LEAVE SUBROUTINE ALTOGETHER
CHARAC, 316           /INITIAL VALUE OF IRI
        325           /N
        315           /U
        332           /M
        256           /B
        275           /.
                      /=
M6,     -6            /CHARACTER COUNT
COUNT,  0             /CHARACTER COUNTER
IRI=10

TYPE,   0             /TYPE SUBROUTINE
        TSF           /PRINTER FLAG UP?
        JMP .-1       /NO: CHECK AGAIN
        TLS           /YES: PRINT A CHARACTER
        CLA           /CLEAR THE ACCUMULATOR
        JMP I TYPE    /AND RETURN

LISN,   0             /LISN SUBROUTINE
        KSF           /KEYBOARD FLAG RAISED?
        JMP .-1       /NO:CHECK AGAIN
        KRB           /YES: READ CHARACTER
        JMP LISN      /AND RETURN


CRLF,   0             /CRLF SUBROUTINE
        CLA CLL       /CLEAR ACCUMULATOR AND LINK
        TLS           /RAISE PRINTER FLAG
        TAD K215      /GET ASCII CARRIAGE RETURN
        JMS TYPE      /PRINT IT
        TAD K2.2      /GET ASCII LINE FEED
        JMS TYPE      /PRINT IT
        JMP I CRLF    /BACK TO MAINLINE
K215,   215          /ASCII CARRIAGE RETURN
K212,   212          /ASCII LINE FEED

A,      0             /A SUBROUTINE
A1,     CLA CLL       /CLEAR ACCUMULATOR AND LINK
        TLS           /RAISE PRINTER FLAG
        TAD K240      /GET ASCII SPACE
        JMS TYPE      /PRINT IT
        TAD K301      /GET ASCII A
        JMS TYPE      /PRINT IT
        TAD K272      /GET ASCII COLON
        JMS TYPE      /PRINT IT
        JMS LISN      /GET CHARACTER INPUT
        DCA TEMPY     /SAVE IT IN TEMPORARY LOCATION
        TAD TEMPY     /NOW THAT IT'S STORED SAFELY,
        JMS TYPE      /ACKNOWLEDGE IT
        JMS CONVRT    /GO TO OCTAL-MAKING SUBROUTINE, THEN
        JMS COMPAR    /CHECK NUMBER FOR VALIDITY AND ACT
        CLA CLL       /THEN, MAKE A ZERO
        DCA SHAZAM    /AND ASSIGN ITS VALUE TO SHAZAM
        JMP I A       /AND RETURN TO MAINLINE
K240,   240          /ASCII SPACE
K301,   331          /ASCII A
K272,   272          /ASCII COLON
TEMPY,  0            /LOCATION FOR TEMPORARY STORAGE

B,      0             /B SUBROUTINE
B1,     CLA CLL       /CLEAR ACCUMULATOR AND LINK
        TLS           /HOIST PRINTER FLAG
        TAD K240      /GET ASCII SPACE
        JMS TYPE      /PRINT IT
        TAD K302      /GET ASCII B
        JMS TYPE      /PRINT IT
        TAD K272      /GET ASCII COLON
        JMS TYPE      /PRINT IT
        JMS LISN      /GET CHARACTER INPUT
        DCA TEMPY     /SAVE IT IN TEMPORARY LOCATION
        TAD TEMPY     /NOW THAT IT'S STORED SAFELY,
        JMS TYPE      /ACKNOWLEDGE IT
        JMS CONVRT    /GO TO OCTAL-MAKING SUBROUTINE
        JMS COMPAR    /AND CHECK NUMBER FOR VALIDITY AND ACT
        JMP I B       /AND RETURN TO MAINLINE
K302,   302          /ASCII B

COMPAR, 0             /SUBROUTINE TO DETERMINE ACCEPTABILITY OF ENTR
        CLA CLL       /CLEAR LINK AND ACCUMULATOR
        TAD TEMPY     /LET'S LOOK AT OUR ENTRY
        SPA           /IS IT A NEGATIVE NUMBER?
        JMP TILT      /YES: THAT'S A NO-NO; TRY AGAIN
        CIA           /NOT NEGATIVE: GET ITS COMPLEMENT
        TAD UPLIM     /ADD 7 TO COMPLEMENT
        CLL           /JUST FOR LUCK, CLEAR LINK
        SPA           /STILL NEGATIVE?
        JMP TILT      /YES: THAT'S ALSO A NO-NO; TRY AGAIN
UPLIM,  7             /BACK TO MAINLINE IF ALL TESTS PASSED
        JMP I COMPAR
```

FIGURE 7.18 Part I Assembly Language Listing

```
        *553
CONVRT, 0
        CLA CLL         /ASCII TO OCTAL CONVERSION
        TAD TEMPY       /CLEAR ACCUMULATOR AND LINK
        TAD M263        /GET TEMPORARY ASCII CHARACTER
                        /ADD FACTOR TO CONVERT IT TO OCTAL
        CLL             /JUST FOR LUCK, CLEAR LINK
        DCA TEMPY       /AND PUT THE NEW NUMBER BACK IN THE OLD SPOT
        JMP I CONVRT    /AND RETURN FROM SUBROUTINE

        *433
TILT,   0
        CLA CLL         /INCORRECT ENTRY SUBROUTINE
                        /CLEAR ACCUMULATOR AND LINK
        TLS             /RAISE PRINTER FLAG
        TAD K277        /GET ASCII ?
        JMS TYPE        /PRINT IT
        CLA CLL         /CLEAR ACUMULATOR AND LINK
        DCA TEMPY       /AND DEPOSIT ZERO IN TEMPY LOCATION
        JMS CRLF        /DO CARRIAGE RETURN-LINE FEED OPERATION
        JMP DIR         /TRY AGAIN
K277,   277            /ASCII ?

STATE,  0
        CLA CLL         /IS LARGER SUBROUTINE
                        /THEN CLEAR ACCUMULATOR AND LINK
        TLS             /HOIST PRINTER FLAG
        TAD K243        /GET AN ASCII SPACE
        JMS TYPE        /TYPE IT
        TAD K311        /GET ASCII I
        JMS TYPE        /PRINT IT
        TAD K123        /GET ASCII S
        JMS TYPE        /PRINT IT
        TAD K243        /GET ASCII SPACE
        JMS TYPE        /PRINT IT
        TAD K302        /GET ASCII B
        JMS TYPE        /PRINT IT
        TAD K311        /GET ASCII I
        JMS TYPE        /PRINT IT
        TAD K337        /GET ASCII G
        JMS TYPE        /PRINT IT
        TAD K337        /GET ASCII G
        JMS TYPE        /PRINT IT
        TAD K305        /GET ASCII E
        JMS TYPE        /PRINT IT
        TAD K322        /GET ASCII R
        JMS TYPE        /PRINT IT
        JMS CRLF        /EXECUTE CARRIAGE RETURN AND LINE FEED
        TAD K252        /GET ASCII ASTERISK
        JMS TYPE        /PRINT IT
        JMP I STATE     /BACK TO MAINLINE
K311,   311            /ASCII I
K337,   337            /ASCII G
K323,   323            /ASCII S
K305,   335            /ASCII E
K322,   322            /ASCII R

AO,     CLA CLL         /A OUTPUT;CLEAR ACCUMULATOR AND LINK
        JMS CRLF        /EXECUTE CARRIAGE RETURN AND LINE FEED
        TLS             /HOIST PRINTER FLAG
        TAD K301        /GET ASCII A
        JMS_TYPE        /PRINT IT
        JMS STATE       /ACKNOWLEDGE IT AS BIGGER NUMBER
        CLA CLL         /THEN CLEAR ACCUMULATOR AND LINK
        TAD NUMA        /DEPOSIT A,
        HLT             /AND STOP


BO,     CLA CLL         /CLEAR ACCUMULATOR AND LINK
        JMS CRLF        /EXECUTE CARRIAGE RETURN AND LINE FEED
        TLS             /HOIST PRINTER FLAG
        TAD K302        /GET ASCII B
        JMS TYPE        /PRINT IT
        JMS STATE       /ACKNOWLEDGE IT BIGGER
        CLA CLL         /THEN CLEAR ACCUMULATOR AND LINK
        TAD NUMB        /DEPOSIT NUMBER B
        HLT             /AND STOP

EO,     CLA CLL         /CLEAR ACCUMULATOR AND LINK
        JMS CRLF        /EXECUTE A CARRIAGE RETURN AND LINE FEED
        TLS             /HOIST PRINTER FLAG
        TAD K316        /GET ASCII N
        JMS TYPE        /PRINT IT
        TAD K305        /GET ASCII E
        JMS TYPE        /PRINT IT
        TAD K311        /GET ASCII I
        JMS TYPE        /PRINT IT
        TAD K324        /GET ASCII T
        JMS TYPE        /PRINT IT
        TAD K310        /GET ASCII H
        JMS TYPE        /PRINT IT
        TAD K305        /GET ASCII E
        JMS TYPE        /PRINT IT
        TAD K322        /GET ASCII R
        JMS TYPE        /PRINT IT
        JMS STATE       /ACKNOWLEDGE EQUALITY
        CLA CLL         /CLEAR LINK AND ACCUMULATOR
        TAD NUMA        /DEPOSIT A IN ACCUMULATOR
        HLT             /AND STOP
K316,   316            /ASCII N
K324,   324            /ASCII T
K310,   310            /ASCII H

SHAZAM, -4             /INITIAL VALUE OF SHAZAM

DIR,    CLA CLL         /BEGINNING OF DIR SUBROUTINE;CLEAR ACC. AND LINK
        TAD SHAZAM      /GET SHAZAM
        SZA             /IF ITS VALUE IS NOT ZERO,
        JMP AI          /GO TO AI
        JMP BI          /OTHERWISE, GO TO BI

M260,   7520           /NUMBER CONVERTER
K252,   252            /ASCII *
$
```

FIGURE 7.19 Part II ...

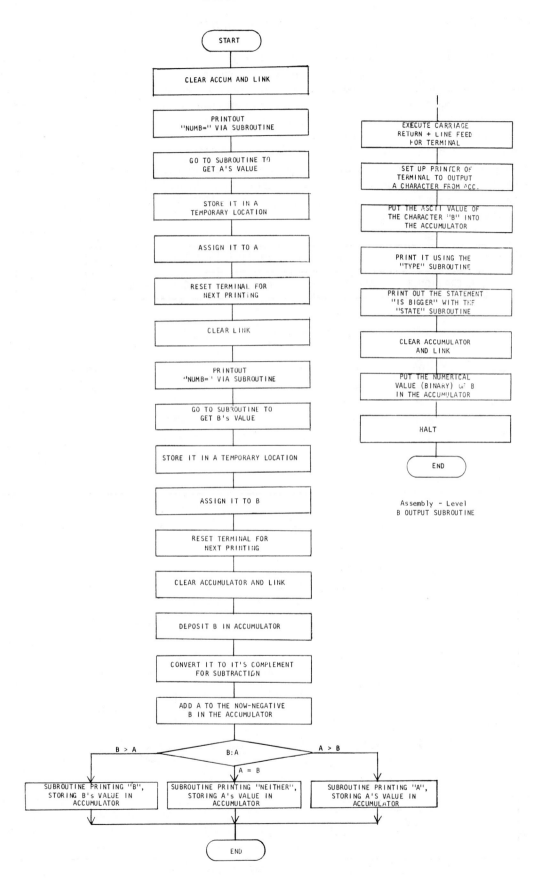

FIGURE 7.19 Assembly-Level Program for Larger of Two in FOCAL

ADDITION OF TWO NUMBERS – MICROPROGRAMMING

Microprogramming, while similar to software programming, requires more detailed knowledge of machine hardware characteristics, and timing must be considered in selecting sequences of microinstructions. The microinstruction is the means of assembling and disassembling hardware elements, functions, or subsystems to perform a processing operation. Speed is obtained in operating the hardware elements in parallel. The larger the word length or number of fields ("micro orders") the greater the number of hardware elements in operation per instruction or the more versatile the microprocessing operation.

The operation $A + B \rightarrow C$ is flowcharted for a Hewlett-Packard 21MX Series minicomputer. The programmer has control of the processor by directing data from all 19 registers and two accumulators onto the S-Bus of the processor and into the arithmetic logic unit (ALU) or memory. From the ALU, the programmer directs his data into the registers along the T-Bus. A simplified logic diagram of the HP 21MX Minicomputer is given in Figure 7.20, and the corresponding control fields are given:

23 22 21 20	19 18 17 16 15	14 13 12 11 10	9 8 7 6 5	4 3 2 1 0
OPERATION	ALU	S-BUS	STORE	SPECIAL

The microprogram, code, and comments are shown here.

Label	OP	ALU	S	ST	SP	Comments
ADD	READ	PASS	S1	M	NOP	Fetch operand
	NOP	PASS	A	L	NOP	Load L with A
	NOP	ADD	T	A	RTN	Add A to operand

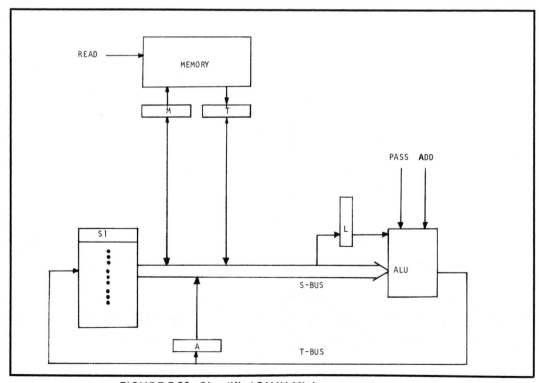

FIGURE 7.20 Simplified 21MX Minicomputer

The three microinstructions in the example are to add the contents of a memory location to the contents of the A-register. Three flowcharts are drawn and scaled in time units (nanoseconds). Each instruction cycle is 325 nanoseconds and the output symbol edge is used to reference the operation completion using the time scale (e.g., all matrix decoding operations of the microinstruction fields are completed by 50 nanoseconds). The height* of each symbol <u>does not</u> imply the duration of each operation. However, once an operation has been completed, the next operation can commence. The various completion times for the same function are dependent on the internal timing generator that is not available to the programmer.

For the purpose here, the address of the memory location is presently stored in Scratch Pad 1. The first step in the execution is to ask the memory to fetch the address operand. First, the S-Bus field is decoded and the operand address in S1 is placed onto the S-Bus (e.g., the operation is completed within 110ns). The store field is decoded and data on the S-Bus are stored in the memory register. The parallel symbol represents an ANDING operating. The digital processing elements for ANDING, ORING, and INVERTING operations are not shown on the logic diagram. The S1 contents stored in the memory register are completed by 270ns. The operation field (READ) is decoded and a read memory cycle is initiated. At this time, the microinstruction is now considered complete. The ALU is decoded and the S1 contents are placed on the T-Bus. The operation is performed but its execution accomplishes no data processing operation. This fetch operation is flowcharted in Figure 7.21.

In the HP 21MX series minicomputers, a read memory cycle will be completed in the time it takes to execute two instructions. Upon completion, the T-Register will hold the data resulting from the read cycle; the memory and the T-Register are not available for the moment. Therefore, a microinstruction may now be coded while the memory read operation is in progress. During the load L with A operation, the contents of the A-Register are placed on the S-Bus (65ns) and are stored into the L-Register (325ns). At this time the fetched information from memory initiated in the previous cycle is now available in the T-Register. The A to S-Bus and S-Bus to T-Bus operatons are superfluous as flowcharted in Figure 7.22.

The third step is to add the two operands (A-Register and the fetched operand from memory) and deposit the result in the A-Register (see Figure 7.23). This is done by placing the contents of the T-Register on the S-Bus (175ns) and coding an add command in the ALU field. The contents of the A-Register, now in the L-Register, are added to the data on the S-Bus. The sum (ALU) output is placed on the T-Bus and stored in the A-Register. The return (RTN) is coded in the special field signifying the end of execution of the routine and a jump to the instruction fetch is made.

SOFTWARE SYSTEM DESIGN

Flowcharting is applied to defining, specifying and designing data processing systems. Its capacity to transcend the hardware and software technology is adequately demonstrated in this section. Flowcharting integrates both technologies and enables communication among the system analyst, programmer, and hardware designer. The examples described here are actual system designs. The system concepts and system details were initially presented using flowcharts. Once agreements were accomplished among the interested parties concerned and specifications were written, flowcharts were drawn embodying these details to delineate system operations to the software and hardware design departments.

The input/output data processing operation is the most sophisticated of the hybrid data processing areas. Hardware (equipment), programming logic, and coding (labeling and identifying) are blended harmoniously to obtain a smooth and reliable operating system. The flowchart information is used to reduce the I/O complexity for the programmer who configures on a local basis or who uses a semi-autonomous execution called an I/O Processor. Coding is written in cooperation with the hardware designer to identify and supply data for transfer and to initiate the transfer operation.

The total microscopic description of a complete I/O operation or data transaction would fill a large book. For description purposes, the I/O is reviewed on a system analysis basis by flowcharting a segment of an I/O operation within a large real time data processing operation (Figure 7.24). Sufficient details are established to enable the reader to understand the I/O operation in terms of hardware and the division of data transfer and data control.

Generally, the computers are off-the-shelf items, and the interface equipment is designed and built to be compatible with the computer and the peripheral equipment (or devices). In the design process, the interface equipment

---

*The height of the symbol has been used by the writer to represent the time duration of a process.

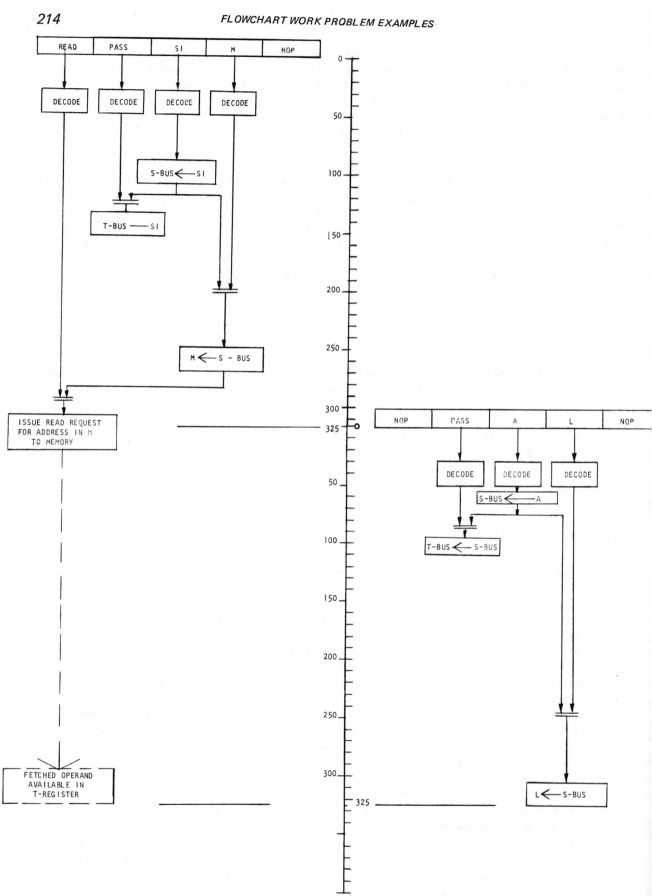

FIGURE 7.21   Fetch Operand

FIGURE 7.22   Load L with A

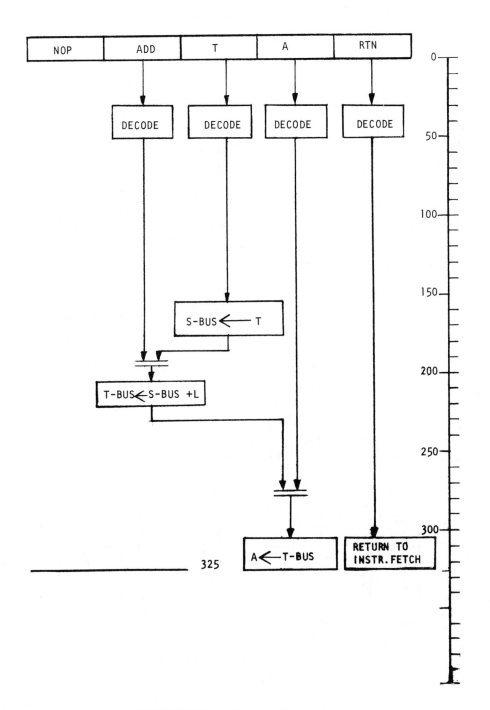

FIGURE 7.23 Add A to Operand

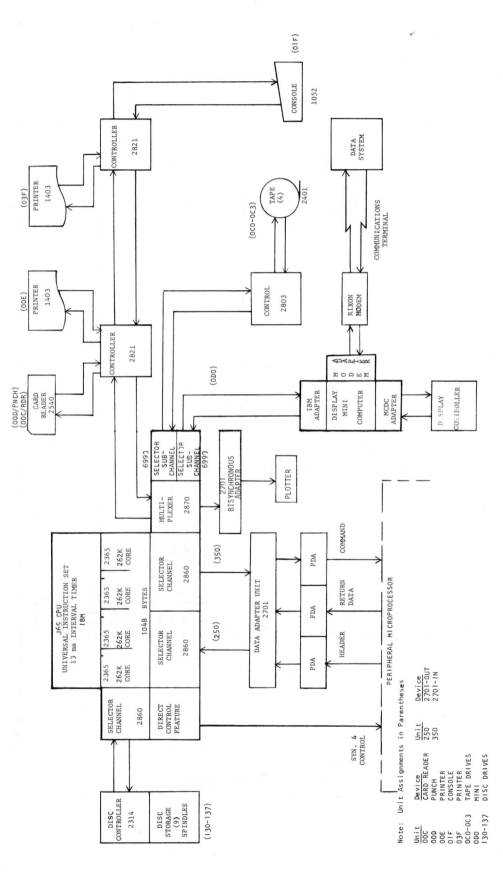

FIGURE 7.24 Real Time Data Processing System

costs are minimized, but some restraints are imposed such as software development and real time operation. The mix of hardware (equipment) and software (programming) costs is found to continually evolve during the development and design phase of a project.

Many of the system details are not present. If sufficient system details were presented in the next few sections, the reader could evaluate the system performance and probably suggest alternate solutions. However, the reader must learn to accept the fact that in many cases, the system requirements can be rigid even to the extent of assignment of specific equipment items. As a matter of fact, there are many real life situations where the system concepts are finalized and the hardware is specified and built, and finally the programmer is told to "make it work." The examples chosen here are firm design requirements and they are imposed on the programmer to supply solutions in the implementation phase of the project or after the system is finalized or built. Some design shortcomings become apparent, but all agree (except the software department) that the simplest solution is a software solution. The examples presented are not a description or evaluation of the system, but flowcharts drawn to express the system operation and to familiarize the programmer with the software requirements and seek his assistance in solving system problems without additional hardware design and other involved project costs. Additional memory is made available (reduce "surplus" core storage) and the programmer is expected to solve all problems with additional programming steps. This is not always the case. The next few examples employ flowcharts to express system operations to the programmer so that he can write the software that is compatible. In effect, they are software specifications.

MAIN COMPUTER I/O INTERFACE OPERATION

This section details an I/O operation between an IBM 360/65 and a peripheral microprocessor (PMP) processing data across an interface in one direction. The block word transfer between the IBM 360/65 and its Peripheral Microprocessor (PMP) is examined and the resynchronization of the index register of both systems is accomplished by software. The physical interface connections involved are as shown. The timing signal bus operations are not detailed here.

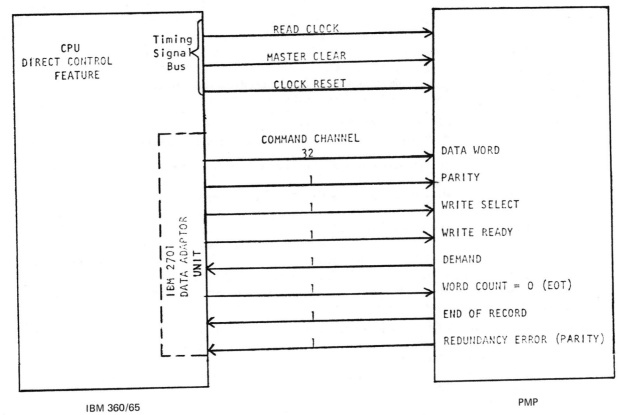

IBM 360/65 and PERIPHERAL MICROPROCESSOR I/O BUS CONNECTIONS

From a programmer's point of view, the flowchart (Figure 7.25) is analyzed in terms of developing compatible software. The flowchart serves as a specification and it serves to describe the system operation. The operation, is the transfer of data from the IBM 360 to the PMP (Box 1 in Figure 7.25). The IBM 360 raises a WRITE SELECT, puts data on the output bus, and raises the WRITE READY (Boxes 2, 3, and 4). The division of operation between the IBM 360 and the PMP is shown by a dashed line in Figure 7.25. The parallel symbol is used to show the simultaneous processing of the command channel bus (32 parallel lines) and the direct EOT line to the PMP. By definition, an EOT can be sent at anytime by the IBM 360 to the PMP.

The command channel bus is a one-way-simplex-transmission system;* see IBM 360/65 and PMP I/O Bus Connections. Once addressed (WRITE SELECT) the transfer operation is a WRITE READY by the IBM 360 and a DEMAND by the PMP. If the PMP is busy or cannot accept the word on the channel bus the PMP responds by not raising the DEMAND signal.** A no DEMAND in return to a WRITE READY means that the PMP has not accepted the data content on the 32 parallel lines. The busy or buffer full operation is shown as Boxes 5 and 6.

For the present I/O system operation, the message word count field (1st word) contains the block or frame transfer count for total quantity of commands issued by the IBM 360 to the PMP for a single block transfer.

FIRST WORD

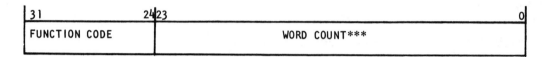

31 24	23 0
FUNCTION CODE	WORD COUNT***

The interface transfers between the IBM 360 and the PMP must always track or be synchronous with each transferred word. This is accomplished in a two step manner. The PMP extracts the transfer count using the function code to determine the first word in a block transfer and stores the value in an index register (Boxes 5, 7, and 8). After the first word transfer, subsequent word transfers (Boxes 11 through 14) are processed and the index register content is decremented by one for each 32 bit word transfer to the PMP from the IBM 360. When the IBM 360 has completed a block transfer, it issues an EOT (Box 15). Upon receipt of an EOT, the PMP checks the word count (index register content) for the value of zero (Box 16) and returns an END of RECORD (Box 20) to the IBM 360 to signify the block count is zero in response to the EOT. The IBM 360 drops the WRITE SELECT (Box 21) and the block transfer is completed (Box 22).

There is always the possibility during the process of transferring words that PMP may test its index register value at the time of an EOT and find $WC \neq 0$. Two situations are possible:

1. EOT is issued by the IBM 360 and the PMP Word Count > 0
2. Word Count $= 0$ in the PMP (Box 11) and the IBM 360 has not sent an EOT (PMP word count < 0)

All that can be gleaned from the above situation is that the I/O transfer operation was not successful, excluding parity reporting.

Once the PMP tests for $WC = 0$ fails, there is a resynchronization or recovery process necessary for both conditions, ($WC < 0$ and $WC > 0$). The $WC > 0$ recovery process is a simple one. When the EOT is received by the PMP and the index counter value is greater than zero ($WC > 0$), the PMP sets a status bit (Box 17), generates an error header (Box 18), dumps the current transfer (Box 19), and issues an EOR (Box 20) in response to the IBM 360's EOT. From the IBM 360 transfer operation, the software transfer completion process is normal with one exception. The status word contains the data informing the IBM 360 that the current completed transfer operation was unsuccessful. The programmer develops his software program accordingly using the I/O status word to determine that a transfer error occurred.

*IBM 2701 Data Adapter Unit, File No. 5360-19, GA22-6844-3, page 2-39 and etc., timing diagram — page 3-15.

**Actually the PMP holds the DEMAND up after the word transfer that filled its input buffer. The IBM 360 cannot raise the WRITE READY with the DEMAND up.

***The first word contains the transfer count for the current block of data including the first word. The peripheral microprocessor word is 64 bits in length. The PMP word is equal to two interface transfer words.

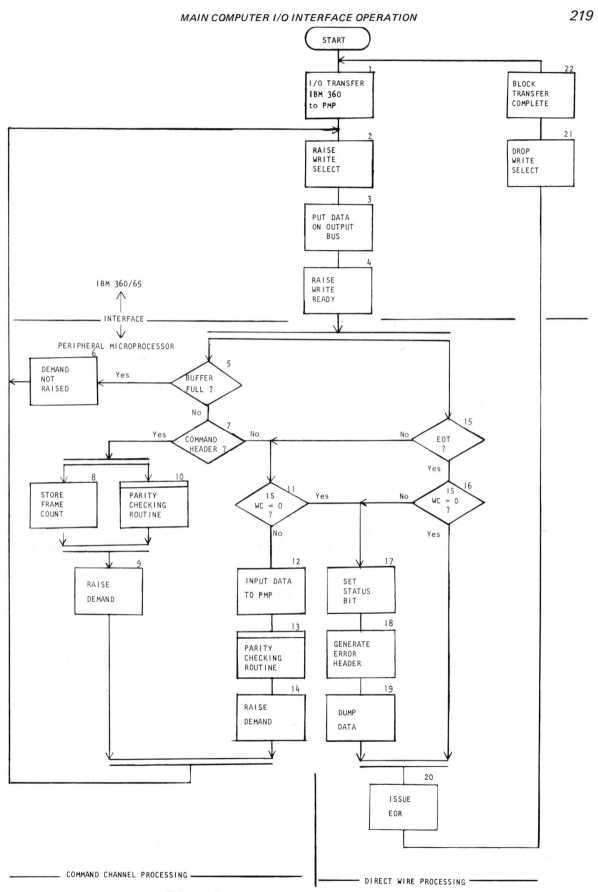

FIGURE 7.25 Command Frame Transfer Operation

In the event the PMP decrements its index counter and arrives at a value of zero (WC $<$ 0) before an EOT is issued by the IBM 360, there are two basic methods of recovery:

1. Permit the IBM 360 to continue transferring words across the interface to reduce its index register to zero.
2. Permit the PMP to issue an EOR (without an EOT being sent by the IBM 360 to the PMP).

The second method is chosen. By permitting the PMP to issue the EOR, the hardware complexity of the PMP is reduced and system response is faster under this condition. Consider the consequences of Item 1 under the previous operating conditions. Suppose the IBM 360 ignored the EOR sent by the PMP (Item 2 above). The software is responsive to processing an EOR to complete a transfer. If Item 1 is invoked, a special case would be necessary to handle this situation. Furthermore, continuing the transfer process can cause problems in the PMP index register operation, since it is set up to use the WC = 0 sentinel to generate an EOR. Since the transfer is unsuccessful, continuing the transfer operation is wasteful. Other problems could develop. If the IBM 360 continues to unload the remaining words in its output buffer and issues its EOT for the current block transfer, a double error may be reported. Recapitulating, this is what is likely to happen. When the PMP word count reaches zero (Box 11) prior to an EOT from the IBM 360, the PMP sets a status bit (Box 17), generates an error header (Box 18), dumps the data (Box 19), issues an EOR (Box 20) and gets ready for the next block of data. There are a certain amount of housekeeping details performed by the PMP to establish its equipment for the next block transfer operation that is initiated by the EOR. Assuming the IBM 360 ignores the PMP's EOR and continues to unload its output buffer, an EOT will be issued when the IBM 360's output buffer is empty (EOT). The PMP will process the EOT by testing its index counter for a numerical value. Because no message count (first word function code) was received from the IBM 360 since the last EOR, the index counter value may be any value (excluding zero for the discussion here). The PMP will process the remaining portion of the total block transfer as an error condition and repeat the previous operation for recovery.

The solution to maintaining word transfer tracking between the IBM 360 and PMP is an EOR processing arrangement that saves time, hardware complexity (PMP), and simplifies software design (Item 2 above). The programmer writes his code to process the PMP's EOR as the sentinel to terminate I/O transfers and tests the I/O status word content to detect any transfer errors. The programmer is at liberty to develop any number of returns to handle the situation of an unsuccessful I/O transfer. Yet, the I/O software is normalized (or generalized) to terminate the I/O operations in response to an EOR and the I/O status word is processed to invoke the appropriate software error return to satisfy system operation.

In summary, the flowchart details the interdependence of the physical connection between two subsystems. The programmer must write his coding to be compatible with the peripheral microprocessor, the interface adapter (byte to a 4 byte word and 8 byte interface word) and the main computer (IBM 360/65). The interface from the IBM 360 to the PMP has the control word, command words, and data words in this sequence over the channel bus. The reverse direction (Figure 7.24) has the command and control words on a separate channel bus (header) from the data words (return data). The two return channel buses contain related data and control for high speed data processing.

MINICOMPUTER I/O INTERFACE OPERATION

The format of the preceding section is continued in this section. The hardware complexity is reduced by having the software perform some of the functions that could be assigned to the hardware category. Again, the computer is the off-the-shelf item and the peripheral devices and equipment are special purpose (custom made) equipment. A device controller is used to integrate a minicomputer (General Automation 16/60) with four off-the-shelf CRT displays. In some cases the first system package description from the system analyst may not be complete or offer alternate solutions.

One of the specifications questioned by the programmer was the polling frequencies. The programmer, attempting to reduce some time-consuming operations, wanted the 40 polls per second reduced to 30 polls per second. The reason given was the CRT screen flicker rate of 30 Hz per second was not apparent to the CRT operator. The system analyst said 40 was necessary to maintain a continuous line drawing capability on the CRT screen. The system analyst's requirements and the programmer's modifications are tabulated.

	SYSTEM REQUIREMENTS			PROGRAMMERS RECOMMENDATION
NO. OF FUNCTIONS PER POLL	POLL FREQUENCY (SEC)	TOTAL PER SEC	FUNCTION	POLL FREQUENCY (SEC)
5	10	50	Monitor levels & slow changing signals	10
5	15	75	Keyboard Operation	20
10	40	400	Line Drawing	40

The programmer raised the polling rate from 15 to 20 per second to simplify his software (see chapter problem). Other system details are introduced as necessary.

The device controller together with the four CRT displays will be called the Manual Control Display Console (MCDC). The physical interface connections with the computer are detailed in the diagram.

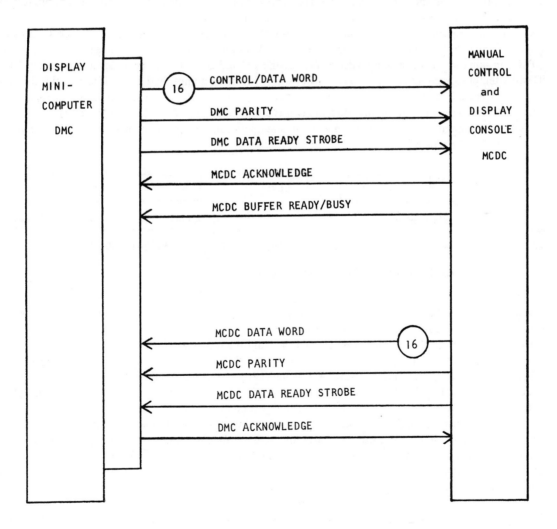

DISPLAY MINICOMPUTER AND MCDC I/O BUS CONNECTIONS

In this example, the programmer is given considerable liberty in handling steady state and transient (or random) system errors. The interface between the minicomputer and the peripheral device controller (MCDC) is considerably less complex than the preceding example, permitting a bidirectional interface description. It should be apparent to the reader that the lack of sophistication places an additional burden on the programmer to develop adequate software.

Normally, the MCDC is always available (ready) and the display minicomputer (DMC) is always in control. The non-poll command transfer is a one way or open loop transfer from the DMC to the MCDC. The DMC puts data on the output bus (Control/Data Word) to the MCDC (Box 1 in Figure 7.26), and generates a Data Ready Strobe (Box 2). The MCDC strobes the control/data word lines and examines the function code for non-poll or poll operations (Box 3).

Assuming a normal non-poll operation (Box 4), the MCDC will respond with an acknowledge (Box 5) that is received by the DMC and the transfer is completed. If the MCDC acknowledge is lacking, the following paths are taken. The MCDC Buffer Ready/Busy is tested by the DMC for a logic [1] (Box 7). If the test proves true, the DMC is permitted to address any remaining MCDC device except the one proven to be busy, (Box 8). If the MCDC Buffer Ready/Busy is a logic [0] and the MCDC Acknowledge is a logic [0], no transfer has occurred (or implemented) nor can it be implied the cause of the problem (e.g., parity checking by the MCDC is ignored for the moment). Since the interface capabilities are limited and monitoring a minimum of MCDC hardware complexity, the programmer must devise an appropriate software method to solve the problem. The approach taken is to determine if a transient error (random) occurred during the process of transferring data or if there is present a steady state error (equipment malfunction or fault). Both are considered in the flowchart. The DMC software may attempt to accomplish the transfer to overcome a transient condition (noise). After N attempts (Boxes 9 and 10), the DMC can skip the transfer attempt (Boxes 11, 12, and 13) and go on to another program operation. Later, the DMC will attempt to accomplish a previous transfer where N series attempts had failed. If successful, the failed attempts are cleared (Box 6). If not successful, a series of M (Boxes 12 and 13) discontinuous transfer attempts are made at N times each before an intermittent error is classified as a steady state error (Box 14).

At Box 3, the function code could be a poll command operation. The poll command is a two step transfer and a two level addressing scheme:

POLL COMMAND	LEVEL	TWO LEVEL ADDRESSING
POLL MCDC	1ST	CRT ADDRESS (one out of four)
MCDC RETURN RESPONSE	2ND	ADDRESS CRT DATA SOURCE

The poll command is a closed loop operation.

The poll command (Boxes 1, 2, 3, 15, 16) begins in a similar manner as the non-poll operation (Boxes 1, 2, 3, 4, 5) in transferring data from the DMC to the MCDC.

Normally, the DMC poll data transfer is accomplished and the next sequential transfer from the MCDC to the DMC must be completed or the poll command operation is null and void. Assuming the poll command is received properly by the MCDC (Box 16) and the MCDC transfers (Box 17) its data to the DMC (Box 18), the parity is checked out on the DMC side (Box 19). If the parity tests checks out true (Box 20), the DMC sends an acknowledge (Box 21) back to the MCDC to verify a completed poll command transfer operation. If the parity check proved to be an incorrect transfer, a software programming choice is shown in the flowchart (Box 24). The DMC could attempt a sequence of N followed by M attempts to determine if a transient error condition or a steady state error condition is present. Going back to Box 16 and then to Box 26, if no acknowledge is received from the MCDC, the DMC software performs the same error testing procedures (Boxes 9 and 12) previously detailed for the non-poll command.

In this example, a bidirectional data flow is depicted and the operations of the DMC and MCDC are interspersed. By being a sequential operation, there is no confusion as to which subsystem (DMC or MCDC) is performing an operation. There is sufficient information to enable the reader to understand the I/O operation concerned with the principles described here:

1. Polling
2. Random and steady state error conditions

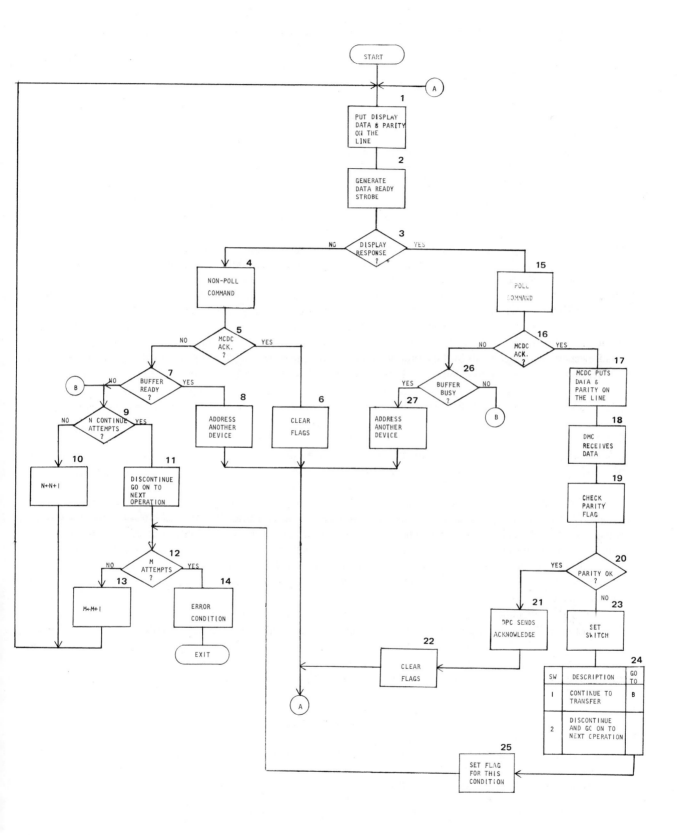

FIGURE 7.26   Minicomputer I/O Interface Operation

No doubt the programmer, with this flowchart system specification operation, could develop the appropriate software to integrate the minicomputer (DMC) with the device controller (MCDC).

INTERRUPT ANALYSIS

The process of program interrupt is discussed on a general basis. The situation is viewed where a series of program interrupts are possible with the potential system problems from a flowchart presentation. The example chosen is a deadlock situation.

A deadlock situation can occur whenever two or more control functions require one resource at the same time. This is best illustrated in an interrupt driven executive control system as shown in Figure 7.27.

In Figure 7.27, the current state symbolically represents the equipment state at any one instance. The process in operation may be a task, an assignment, or the processing of a statement. The normal operation in advancing from one machine state to the next state is shown in heavy flowlines. By definition, the occurrence of an interrupt between the completion of one state (or process) and prior to the next operation is not an interrupt. The interrupt between two boundary states is in essence the servicing of the next sequential state.

The interrupt occurs whenever a process is in operation and its completion is prevented momentarily. For this event, and to re-establish the machine state at this event, all data and status must be stored. The interrupt is then honored and processed. In priority interrupt systems, any low priority interrupts are stored and thus are not really considered interrupts since they do not affect the current process. However, by definition, the interrupt processing operation can be interrupted by a higher priority. Again, for this event, and to re-establish the machine state at this event, all pertinent data and status must be stored. The next interrupt is then honored and processed. As shown in this flowchart, a conflict exists in the nested interrupt processing operation. There is the possibility of alternately interleaving two loops. The problem arises in the STORE CURRENT CONTENTS as related to nested interrupts. For example, suppose control function A has requested and been given control of resource X and attempts to request resource Y. Resource Y has been requested by, and control given to, control function B which is an interrupted state. Control function B is temporarily promoted and, before relinquishing resource Y, requests resource A. This is a deadlock situation in which neither control can proceed. A possible solution of deadlock prevention is given for this example. If both control functions had first requested resource X and then Y, the deadlock would not have occurred. Avoidance of a deadlock situation must be analyzed on an individual basis.

A further problem can exist in this situation for a real time operating system. After all interrupts have been processed for the system, the ESTABLISH STATE PRIOR TO INTERRUPT can cause system oscillation or instability. Time has transpired and the stored data may be stale or corrupted by the total interrupt processes. If a time priority system is imposed on a hierarchy priority system, the previous nested interrupt situation becomes compounded. Obviously, if the former problem (lock out) is solved, the latter problem (instability) remains. The instability problem has many ramifications, especially for the real time operating system. Solutions are not offered, but they might include polling, rollback or roll forward (resynchronization), and redundant operations (not equipment).

The potential deadlock situations arising from priority interrupt queues are quite visible in Figure 7.27. From our previous discussion concerning loops, nested loops, and the execution of nested loops, the loop operation can only be solved or resolved by abiding by the rules of nested loop operation. When a system operation is flowcharted and the result is an interleaving process of two or more loops in a manner similar to the one in Figure 7.27, the software operation must be configured to eliminate the problem area.

REFERENCES

1. Organick, Elliott I. *Computer System Organization – The B5700/6700 Series.* London and New York: Academic Press, 1973.
2. Nassi, I., and Shneiderman, B. "Flowcharting Techniques for Structured Programming." Technical Report No. 8, 18 pp. State University of New York at Stony Brook, 1972, State University of New York at Stony Brook, Department of Computer Science.
3. Dahl, O. J., Dijkstra, E. N., and Hoare, C. A. R. *Structured Programming,* pp. 1-72, London and New York: Academic Press, 1972.

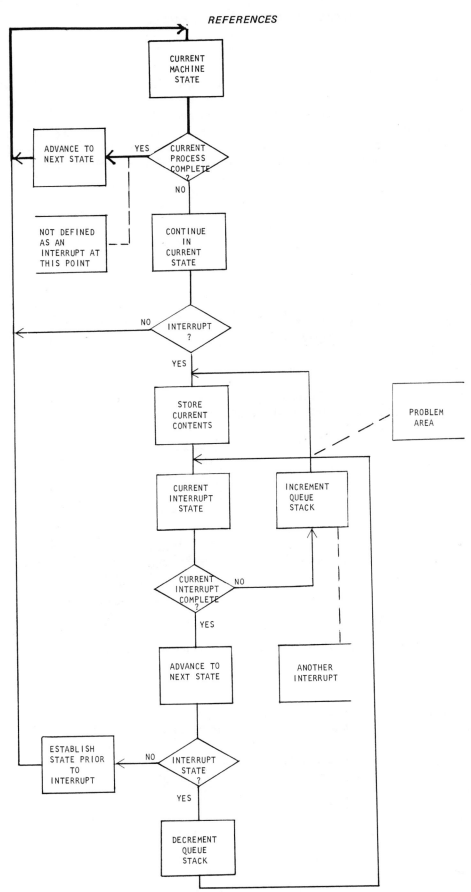

FIGURE 7.27   Interrupt Analysis

PROBLEMS

1. A computer program was developed to provide flexible and dynamic manipulation of new and on-going tasks required for realistic evaluation of operator load. This requirement stems from the real world situation in which the operator continuously balances task requirements temporarily in order to accomplish the mission within time constraints. In other words, if he finds that the imposition of a new task will produce an overload, he delays the new task as long as possible (an allowable delay); however, when the allowable delay time (if any) has expired, he initiates the delayed task and delays an on-going task as long as possible, then shifts the delay to another task, etc. It is only when all on-going tasks have been delayed as long as possible that a clear-cut overload situation exists. Flowchart the basic logic of this narrative.

ANSWER

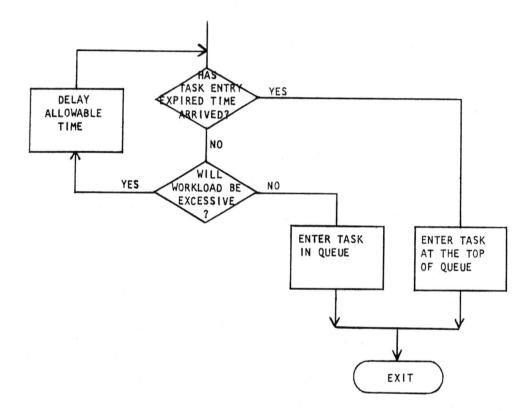

2. The following information was supplied to the programmer from the system analyst:

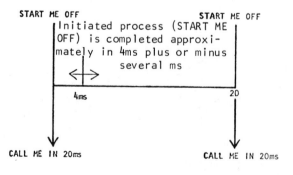

The programmer generated the following flowchart:

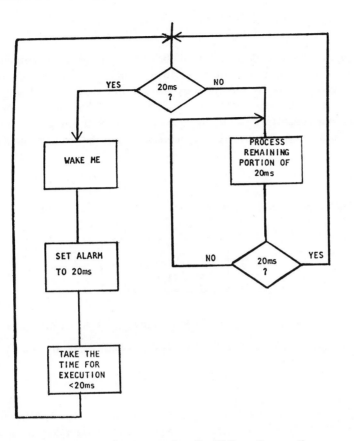

The programmer experienced system instability occasionally. If the call operations exceeded the allotted 4ms, the 16ms period would be squeezed for the remaining time interval and the operations performed for this period of time would suffer. When the called routine exceeded its allotted time of 4ms, the next sequence of operations is performed as scheduled and, if not completed, a backlog is built up. Eventually, the accumulation of the backlog exceeds a time limit where processing the data has no value. The stored data is flushed and a WAIT is invoked to get back into synchronization. The wait period varies in accordance with the flush operation selection and its time duration.

The programmer submitted a software solution that avoids the above mentioned problem at a small cost in system performance. The called routine (same one as before) is initiated and completed. Upon exit (or return), a request for repeat in 20ms is made. The variable time duration will occur between each 20ms rather than within the 20ms period. The initial and modified timing diagram are shown below:

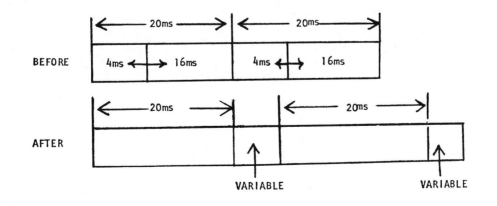

Modify the flowchart to reflect the programmer's recommended suggestion.

ANSWER

Not included.

3. In the I/O minicomputer example, flowchart the polling operation for five operations at 10 per second; five operations at 15 per second, and 10 operations for 40 per second. For the programmer's values of 10, 20 and 40. Hint: For every two 40s do a 20; for every two 20s do a 10. Examine both flowcharts. Does one flowchart appear exceptionally complex? If yes, what about coding the program? Corollary, does one flowchart appear simple and straightforward? If yes, what about coding the program? Hint: Do not draw a very detailed flowchart; just enough to convey your comprehension of the problem and its solution.

ANSWER

Not included.

4. In the COBOL preprint form example, the following flowchart symbol was used:

```
|  | PERFORM     |  |
|  | MAIN-PRINT  |  |
|  | 300 TIMES   |  |
```

MAIN-PRINT is a subroutine called by TEST-FORMS. Flowchart MAIN-PRINT as a nested loop operation.

(Hint: LINESKIP is repeated 22 times and MAIN-PRINT is repeated 300 times.

ANSWER

Not included.

5. If the opportunity is available and with the computer operator assistance, the following exercise should be attempted. Using a COBOL program similar to the preprinted form example, write a program of a few lines of text; Name and Address. Do not advance the output sheet to start position. Align the printer page in any middle page position with the help of the computer operator. Run the program text twice. Lift the cover up on the printer and move the sprocket hole page to align the text material to be located on the output page as defined in your data layout. See how many times the operation is repeated until success is achieved. Then print 10 pages.

ANSWER

No answer required.

8

CURRENT STATE and FUTURE PROSPECTS

Flowcharting continues to proliferate in all fields and the user of data processing in such areas as banking, insurance, manufacturing, the media, medicine, the military, and transporttion will find flowcharts in his descriptive literature. It is difficult to avoid the subject of flowcharting, given the current state and extent of computer technology.

Computer hardware costs remain relatively stable (neglecting inflation), but the cost of software continues to escalate dramatically.[1] Consequently a new approach to software development is being pursued in search of improved efficiency and economy. Programming nowadays is being streamlined, standardized, and structured for production by automated methods. Automated program writing techniques are in the offing.[31]

As presently conceived, automatic program writing proves to be highly dependent on flowcharting[2] (see Appendix A). Computer consoles with interactive flowchart displays permit construction of large programs from "canned" segments by file-retrieval methods. Program writing costs come down when the programmer can assemble his program instead of writing all of it specially for his requirements. The building block concept is conducive to program testing by chaining modules, one at a time, and testing each additional link as it is added.

Graphic displays and their supporting software are becoming so inexpensive that it is only a matter of time until every household will be equipped with some kind of interactive display terminal. Already many department stores use cash registers with CRT (cathode ray tube) displays and pushbutton controls for recording sales transactions. Therefore at home, at the office, at shopping centers, and at school, the interactive display terminal will play a major role in our lives.

The widespread use of various kinds of interactive display terminals means that the public must be educated to understand and operate the device. This is being done in high school and shortly will be introduced at the junior high school level. Every professional will come in contact with some form of digital processing. In order to master the subject matter, flowcharting will be used as an educational tool. Extensive use and application of flowcharting will probably uncover new uses that are not apparent at this time.

Flowcharting is replacing many management software control methods and techniques, because it is more effective or because the other methods have failed to perform their required functions. Flowcharts are now used by management as a control tool to oversee the development of commercial and industrial computer programs. The charts expedite the administrative process of auditing and accounting of program control and they provide the dynamics of an otherwise static presentation of blocks and tabulations.

Flowcharts are used extensively as lecturing aids for graphical presentation and communication. The charts have advanced to the stage of visual pictures in which motion is obtained by sequencing to the next level of increasing detail. Flowcharting is an established technique that can be adapted easily for use in presenting a series of steps in a predetermined order.

Literature and publications are using flowcharts as a prime source of communication. Areas of communication include: reference and text books; research (design, development, and planning); and contract specification and commitment.

AUTOMATIC PROGRAM WRITING

Automatic program writing is the synergistic result of a number of disciplines applied to this goal. A number of fragments are being integrated to construct a simplified automatic program writing system with a total capability still in the distant future. The overall aim of the automatic program writing system is to allow direct entry and use of the computer data base by flowchart addressing and retrieving stored computer programs at any level of detail and complexity. The programmer can develop, modify, and finalize a computer program by assembling and integrating procured segments from the data base to generate a new software program. The area covered here is restricted to flowchart representation of programming language and conversion of chart symbols to machine statements. (See Appendix A.)

Two areas are addressed here:

1. Symbolic representation of one or more computer statements.
2. Concatenation of the symbols to generate a computer program.

A total complex software program must be dissected into definable units before it can be represented by symbol notation. Structured programming is a technique that entails modularity, hierarchy, and conventions to generate a computer program by assembling completed programmer assignments and tasks. Instead of a team of programmers using structured programming to write a new computer program, the process can be automated by addressing the "canned" modules on file and assembling them into a new computer program. A number of examples presented here are related to automatic flowcharting of a program listing.

It is quite simple to represent a program statement in flowchart form. Figure 8.1 illustrates the process for FORTRAN. The use of CALL entails a software hierarchy. An entity is defined in terms of a function, an algorithm, or a process in which the label or name serves as a key to search and retrieve the subordinate program. This method enables the programmer to utilize procedures pre-written either by him or someone else. This is a simplified form of automatic program writing without the actual writing of each program statement by the programmer. The linkage edit utility program used to generate a load module is another form of automatic program writing. The programmer constructs a computer program by assembling a number of modules using a utility program to compile a program whose segments are computer resident (memory storage and disc file) or externally stored on tapes and card decks.

Many routines are designed especially to be general purpose and accommodate a wide range of applications. There are numerous commercial software packages that fall into this category. The vendor has invested in a product for sale. The more flexible and versatile the software package performance, the more users who can be accommodated with a minimum of adjustment for each sale. The means of distribution are very easy (e.g., tapes, discs, tape cartridges, and card decks). There are numerous instances of automatic program writing in fragmented form. All that remains is to construct a file or data base with a mechanized method for retrieving the information. In addition, the file maintenance procedure must permit expansion, modification, addition, and deletion of the

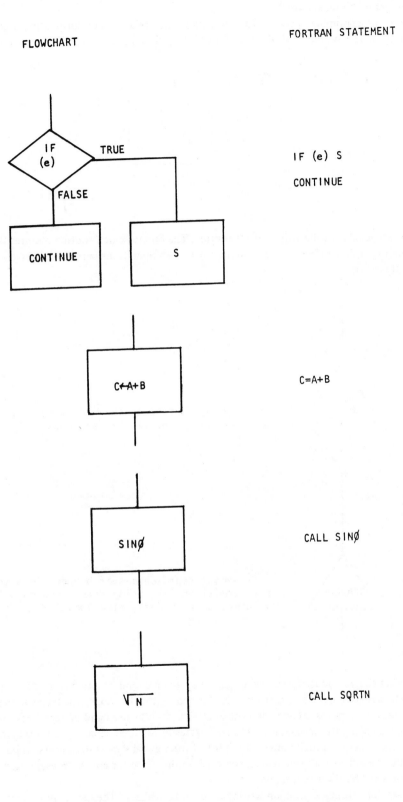

FIGURE 8.1   Conversion of Flowchart Symbols to Program Statements

"canned programs." Flowcharting in conjunction with an interactive display dialogue are major items of an automatic program writing system.[3]

A number of programming operations are used daily and are a form of automatic program writing. Thus, for example, the flowchart symbol with the read input text material calls into action an I/O utility routine.

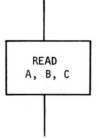

The same can be said for the write output routine. The binary search routine requires that the file list be ordered ascending or descending. The concatenation of two operations permits the programmer to express this software configuration.

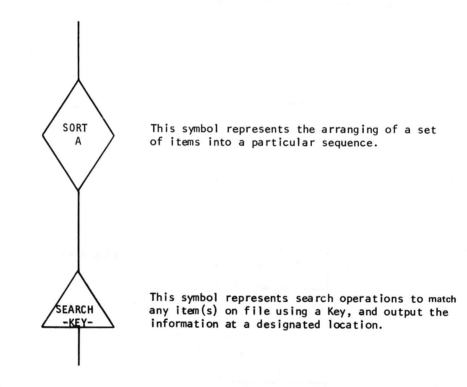

This symbol represents the arranging of a set of items into a particular sequence.

This symbol represents search operations to match any item(s) on file using a Key, and output the information at a designated location.

The reader is being introduced to a new form of key and a complex data base and structure. The key is a combination of a flowchart symbol and text material defined by the dictionary for the automatic program writing system. Together, the input is decoded as a list of attributes to be used and matched against a directory to seek and retrieve stored computer programs in relative degrees of required capabilities (100 percent or less). Normally this is being done when the programmer scans a list of software abstracts to obtain computer programs to assist him. Many of these software abstracts use a standard outline form for concise presentation of the computer program description and hardware configuration.

The full scope of automatic program writing is yet to be realized. The user may be a system analyst who is skilled in problem solving and has a substantial software background, but is not an expert programmer. He can evaluate specifications and express his logical solution in a natural way using flowchart symbols on an interactive CRT display. The solution is obtained by building a flowchart rapidly and the screen area can be rolled indefinitely in all directions . Obvious mistakes are detected and corrected.

Normally, program writing is a series of steps to assemble primitive program structures into more complex structures. The logic is a series of flowchart symbols and text details. The automatic program writing system converts this tree structure into a string of symbols and uses them as a key for searching. Ideally, a perfect match between individual input increments and corresponding program segments is desired, but not always obtained. The computer programs on file are completely cataloged and indexed in terms of their primitive elements as well as the total unit. The search process matches the input string with any string of any program unit on file. Since flowcharting is a programming language and is computer independent, the retrieved information can be displayed in two forms:

1. Program Listing
2. Flowchart Form
   a. Flowchart of the listing
   b. Flowchart of the listing that is independent of the language and computer configuration

In Item 2.b., if the data format is constructed as stated, the input flowchart symbol sequence and the file data can be checked by any checking routine. The two flowcharts (input and Item 2.b.) can be overlayed on the CRT screen and the user can decide the degree of matching. Here program schema and flowchart equivalence techniques are a must. An alternate method would permit the user to examine the program listing and its equivalent language flowchart. If necessary, the user would modify the retrieved program statements or use them as a guide to make a substitution. The process is repeated until a computer program is obtained. Meanwhile, the system utility programs constantly assist the user by checking his logic and supplying error messages as necessary. At this point, the program is syntactically and semantically correct with one more operation yet to be performed: Program checkout with actual problem data.

The user must check the stability and performance of his program synthesis. The program is run against "live data" supplied by the user. Any overflows, conflicts of allocation of resources, missing items on the part of the user, and numerous subtle details are revealed. If the user is clever with his input data, he may check his program still further. Erroneous and pseudo data may be inputted to see if the computer run will hang, stall, or deadlock. If any of these problems do occur, he can modify his program accordingly. Tests are inserted (decision symbols) to detect the occurrence of the malfunction so that a remedial or recovery route can be initiated.

The automatic program writing system is in its infancy. The exposure here has been the use of flowchart symbols as a searching key and a complementary data structure that is compatible for string processing. Top down and bottom up program structuring are necessary. When a complete program unit is not obtainable from the file, top down structuring is necessary to obtain the parts from file to assemble the program. The bottom up structuring is necessary so that the synthesis is responsive to the user. Flowchart equivalence achieved by manual (operator) methods are recorded in memory to serve as a precursor for computer program segments on file. Whenever search requests go unanswered, the data file is updated for the next user.

AUTOMATIC FLOWCHARTING

Automatic flowcharting in some form has existed since the word flowchart was coined. The term is most associated with the flowcharting of a program listing. The flowchart can be a brief or a detailed representation of the program. The level of detail is a function of the automatic flowchart package or program capability.

There are a number of methods for flowchart generation by other than manual means. One approach produces flowcharts by analyzing the program's source code. Flowcharts produced by this method require no additional programmer's effort, but may be at the statement level. A second approach uses a flowchart language or commands. Although programmer effort is required, the level of detail, format, and general style of the flowchart can be controlled. There are two alternatives commonly available with the second approach. A high level language can be devised to control another input source (programming language or machine language) or serve as input since it is compiled and translated into object code directly or by indirect means via another compiling process. This last form of automatic flowcharting relieves the programmer of the tedium of manually drawing the flowchart. The flowchart symbols are stored in the computer for the user in the same manner that such symbols are "stored" on a flowchart plastic template. The handwritten text material is automated by a keyboard accessory. Positioning of the flowchart symbols is automated by a light pen which the operator can use for inputting and editing information on the CRT (cathode ray tube) display. The flowcharting is in essence, a design tool,[4] rather than a documentation package. The problem of illegible handwriting is overcome, and there is little need for manually drawing flowcharts.

From these two basic approaches, there are numerous variations including flowchart dictation[5] to relieve the programmer of drawing flowcharts. Both automatic flowcharting methods greatly reduce the demand made on the programmer and in effect, encourage more comprehensive program documentation. Quality of automatic flowcharting is always uniform and consistently clear and legible. Automatic flowcharting standardizes flowchart drawings and their maintainability.

The need for flowcharts and flowchart documentation as a means of communication cannot be disputed, but hand drawn flowcharts are costly to prepare and maintain. Like handwritten notes, the typewritten or printed page corollary for flowcharts is essential. Various computer programs have been developed and many are offered for a price to mechanize flowcharting. The flowcharting techniques are an evolutionary process probably dating back to Leibnitz and Pascal in the seventeenth century when they were doing their work in designing calculators. Babbage, in the design of his analytical machine, must have drawn sequencing and processing charts.

Impetus in flowcharting began in 1963 and considerable activity in the field has been present ever since. The ground work by A. E. Scott,[6] L. M. Haibt,[7] and D. E. Knuth[8] proved the feasibility and value of automatic flowcharting. In the same year, the essential contents of the Proposed American Standard Flowchart Symbols for information processing were published for concurrence by the industry at large. In 1965, the Proceedings of the Fourth Meeting of UAIDE produced four papers related to flowcharting.[9] A detailed history of automatic flowcharting would be of academic interest only and would not serve any useful purpose here. Literature abounds with numerous flowchart drawing programs.

The quality of these systems varies widely. Types of output units for drawing flowcharts are also varied. Some of the systems use CRTs for producing flowcharts. Others produce flowcharts on printers and by use of plotting devices.

There are a number of sources that evaluate, detail, and list the cost of available commercial flowchart packages.[10-13]

FLOWCHART DICTATION[5]

Flowcharts have been used for "shorthand" notation by programmers and hardware designers for a long time. Now, the task of program coding is being replaced by dictation. The computer program is flowcharted in crude fashion and then refined. Once the flowchart is finalized, the programmer dictates the information, over the phone if necessary, and it is recorded on a magnetic belt recorder. The secretary transcribes the dictation into a typewritten coding form, both human readable and machine readable. It is fed through a page reader to produce punched program cards and listing. The cards and listing are checked by the programmer and then filed. The system operation claims an error rate in program coding of less than 1 point. The high level language is COBOL and the system is IBM.

The flowchart dictation system is quite dependent on the skill of a trained secretary. Another system, the Flowchart Programming System (FPS), briefly detailed in Appendix A, is completely automatic from the word GO. The programmer is relieved of drawing any flowchart symbols. The graphic display is the programmer's sketch pad. Once the flowchart is finalized, he can call and immediately get his flowchart translated into high level language. Any editing is done on-line and a listing and card deck can be made at the push of a button. Admittedly, the first system[5] is less costly than the second one and who wants to be denied the pleasure and pulchritude of a private secretary?

AUTOMATIC DIGITAL HARDWARE DESIGN, MANUFACTURE AND TEST

Flowcharting has long been second nature to computer programmers and system analysts. After the system is defined and designed, the digital hardware is implemented by the logician using the performance specifications of the logical elements on module cards.* Automation has been introduced at this level. Computer programs assigned logical elements to particular module card designs, and wiring diagrams were generated. This is outmoded. The system analyst, just as the software analyst, flowcharts the system specification. He evaluates, modifies, and

---

*"Module cards" are phenolic or fibreglass printed circuit boards on which are mounted aggregations of either discrete components or integrated circuit devices.

rearranges the flowchart design until he is satisfied with it. The flowchart is converted to logic elements in terms of digital modules by automation.[14] The current techniques of assignment of logical elements to cards, card layout, nest wiring, and cabinet wiring are all automated. Now automation is introduced at the beginning of the hardware manufacturing phase.

The prospects for flowcharting in hardware design phase are tremendous, since there is a vast vacuum here. Flowcharting is not intensively used here as it should be, partly because hardware designers have exhibited a tendency to resist change. Circuit design has become automated and equipment design is becoming the automated assembly of module cards. The logic on these cards no longer consists of a few logical elements (NOR, NAND, INPUT EXPANDER OR EXTENDER, etc.). Today a solid state chip comprises 500 to 1000 logical elements or more. Minicomputers can be built using a single printed circuit board plus power supply and memory. Flowcharts offer the means to "walk through" the logic and they can be used to define each functional operation by code correlation[15] down to single bit operation in multilevel flowcharting.

It is difficult to understand the operation of digital logic using text material, timing and block diagrams, charts, lists, coding tables and equations. At design review meetings, the logic designer with his review package of the above items without a flowchart can seldom convey a description of operation and give visibility to his technical solution. His audience in return, through a lack of understanding, cannot evaluate the design nor make a contribution. The way to solve this bottleneck is to use flowcharts as one of the major items in the design review package.

MANAGEMENT CONTROL

Flowcharting allows application of a variety of technical tools which can make a major contribution to the technical management control of a real time software/hardware development program.

The management control method to be outlined here has the following attributes:

1. It is foolproof and error free.
2. It enables technical control of contract commitments and end item implementations.
3. It functions manually initially, with the potential for automation.

First the system specifications are flowcharted to generate the primordial System Specification Flowchart. This flowchart provides the most effective means of depicting the functional interrelationships and interdependence of the elements of the system. Typically this chart is a multilevel graphical presentation that will enable complete functional definition, control, and audit of a large development program. Use of the chart ensures that the program development is logically correct, complete, and consistent with contract commitments and requirements.

The system specification and performance text material are used to generate a system specification flowchart. The level of detail is compatible with the system specification and one of the advantages derived from the SS flowchart is its ability to reveal any inadequacy in the system specification. (The word "system" used here means hardware and software.) Once the SS flowchart is finalized and agreed upon, the logic is verified using a simulation program. If the simulation program is a modest one, then the simulated functions are traced (or inked) on the SS flowchart and the residue are nonsimulated sections.

The simulation is necessary to evaluate and check the proposed logic, system configuration, and to obtain and/ or optimize system parameters. The final SS flowchart is used as a design specification. It is placed under configuration control for the remainder of the program development subject to a review board for each addition, modification, or deletion. It becomes a reference document to the Design (hardware and software) and System Test and Integration departments. It is used for acceptance testing and contract sign off, because it contains the contract requirements in graphical form.

The system specification flowchart is used to generate a system implementation (SI) flowchart. The SI flowcharts are generated by the software and hardware departments and are checked for compatibility using the SS flowchart. The SI flowcharts are used to modularize task specifications for staffing requirements and assignments. Once the task assignments are completed, they are transcribed onto the SI flowcharts for task identification and responsibility.

The SI flowcharts are further detailed to write module specifications prior to coding. The software programs are generated and they can be automatically flowcharted from the source language or object language statements. If an automatic flowcharting package is available, it is used. Otherwise, a hand drawn flowchart is made of the program listing.

There are several methods of evaluating the computer program. Since the SI flowchart (software) is the build for the SS flowchart, it can be used to drive the simulation program. This would require the generation of a new software program that is time consuming and costly. Another method which is effective and recommended is the following approach. The SS flowchart is the system logic and operational details. Regardless of what high level language is used to write the program, there is a one-to-one correspondence between the SS flowchart and the SI flowchart.[32] Program equivalency of two programs is proven when the same input delivers the same output. They are expressed as follows:

Program (Schema)*   S ≡ S′

Two flowcharts are logically the same (execution of S), if the logical flow sequences and branches are the same. They are expressed as follows:

$$F \equiv F'$$

The automatic flowchart of the program statements is generated (or compressed) to a level of detail that is compatible to the SS flowchart (e.g., bottom up parsing). If necessary to obtain flowchart equivalence ($F \equiv F'$), the SS flowchart is further detailed (top down parsing or flowchart grammar production).

The method of flowchart equivalence is obtained by manual means. The computer program flowchart is checked logically with the SS flowchart item-per-item visually by the programmer(s). A one-to-one mapping of logic flow must be obtained. Since the SS flowchart is the contract obligation and it has been used to drive (write) the simulation program, it is the reference document. The computer program flowchart is altered (program change) to obtain equivalence. Once this is done, a high degree of confidence is assured prior to system testing and integration. By definition, the computer program is logically correct ($F \equiv F'$) and operational (simulated system). The hardware counterpart is flowcharted and checked with the SS flowchart for the same results. The end items (computer program and equipment) should not experience any (major) problems during system test and integration.

The test plan is written from the SS flowchart and the computer program is responsive to the SS flowchart. Technically and logically, after corrections, the computer program should run successfully during testing. At acceptance sign off, the test plan is checked against the SS flowchart for contract commitments and the computer program is run using the program test for certification.

A system specification is graphically represented as a flowchart at the beginning of a development program and is used as a reference document throughout the program. Every design, build, and/or end item is flowcharted and checked for equivalence prior to the next program development phase or step. In this manner complete technical control of the program development is ensured in terms of contract commitments and customer response. Any customer changes are drawn onto the system specification flowchart. For any customer change, the system impact is visible, and the system specification flowchart is used for proposal purposes.

FLOWCHART EXTENSIONS

Several flowchart extensions have been devised based on modifications and interpretations of the flowchart symbols. In a way, there are competitive methods as well as substitutes for conventional flowcharts, and there are flowcharting extensions to obtain more versatility and flexibility. The flowchart extension coverage is from a simple outline format, through a continuous flowchart to an assembly of flowchart symbols whose operational description is designated as a flowchart machine.[16] There are many flowchart variations within this range.

FLOW OUTLINING

Flow outlining is one step below flowcharting and possibly one step up from just plain outlining. Flow outlining is a method of listing in sequential order the steps to be taken in which each step is identified by a label. Reported by Gant, this method requires less time for preparation, permits more detailed remarks, and is easier to modify or correct.[17] Gant's method consists of a symbolic identification and a statement of what is to be performed. The

*See Appendix B for formula explanation here and for formula explanations for the remaining sections of this chapter.

symbolic identification is used in coding, thus making it easy to compare the coding with the outline. Statements can be added at any time using convenient subscripts.* The flow direction is top down without any lines for flow direction. Any deviation from this normal flow path is accomplished by labeling. An example of Gant's flow outlining is shown in Figure 8.2.

Two other forms of flow outlining are presented in Figure 8.2. On the right side of Figure 8.2B and Figure 8.2C a flow line is used to obtain a two dimensional effect that cannot be obtained in Gant's presentation. Still, no symbols are used and no templates are required to form flowchart symbols. Although not expressed, the decision symbol or branch operation is very apparent. The advantages expressed by Gant are retained. The text material is neither restricted nor limited as to position location on a page and text volume. Any arrangement of the information is permissible.

The diversification of flowchart application is aptly expressed in the three reference sources on flow outlining in Figure 8.2.

Gant's presentation is on software and conducive to his programming effort. Figure 8.2B is copied from a DEC** publication. The flow outline depicts a hardware operation, and hardware components are identified in the operation.

The illustrations in Figure 8.2C were obtained from a newspaper, "Computerworld." Here, the flow outline (on the left side) is a series of questions and answers on private lines (> 4800 bits per sec) and dial up lines (< 4800 bits per sec). The flow outline directs the user to answer each question and follow the path that is his response to the question. Logically, if the questions evoke the correct response, the correct selection will be made. Similarly the flow outline on the right side of Figure 8.2C enables the user to decide the class of data links and modems he needs.

CONTINUOUS FLOWCHARTS

The continuous flowchart is analogous to a road map — in fact a set of road maps at successively more detailed scales. An operation shown simply as a circle in one chart can be detailed in the next lower level chart. The analogous situation is the state highway map, which shows towns as simple circles, and the individual town maps, which detail the contents of the circles.

In continuous flowcharts, the flow path is a channel presentation with allowable space for detail description and direction indication. All flow paths are continuous except for the start and end points. The continuous flowchart presented by Gieszl[18] is dynamic and dramatic for his simulation application. Three levels of detail are copied from Gieszl's paper to show the continuity concept. The procedure to be followed for this is to circle the area in question; the top section in Figure 8.3. The avenues of entry and egress from the circled area are preserved. Then expand the logic (via a continuous flowchart) inside the circle. This is shown in the middle section of Figure 8.3. The "forces available" have been expanded into three types: helicopter; destroyer; and aircraft. The bottom section is a more detailed view of the helicopter operation. Clearly this process can be continued down to the level of detail to where the simulation can be circled. When labeled and indexed properly, the entire set of flowcharts serves as a complete documentation.

THREE DIMENSIONAL FLOWCHARTS

Normally any flowchart is a two dimensional drawing. The third dimension can be provided by successively more detailed (multi-level) charts such as those described above, or multi-level flowcharting covered in Chapter 3. The third dimension could also be the time parameter, as detailed in Chapter 4. It appears that in the future multi-leveling flowcharts and time order flowcharts will become prevalent.

PENETRATION

Current application of visual and graphical techniques of learning may portend things to come. Many graphic diagrams are in essence flowcharting, but are called by other names. They utilize the basic flowchart rules and

*In Figure 8.2A, statements BB2A and BB2B could be statements added between BB2 and BB3.
**Digital Equipment Corporation

AA   Reset SUM Y to zero, eject page.

BB   Read a card, branch to BBi if i = 1, 2, 3, 4 where i is digit in column 1. If i ≠ 1, 2, 3, 4, go to CC.

BB1   Alphabetic comments card, print columns 11–80 alphabetically, go to BB.

BB2   Data card, containing X_1, X_2, X_3, compute $Y = X_1 + X_2X_3 + X_1X_2$.

BB2A   Print X_1 X_2 X_3 Y.

BB2B   SUM Y + Y → SUM Y, go to BB.

BB3   Total card, double space, print SUM Y, go to AA.

BB4   End of job, start next program.

CC   Error condition, print error message, pass remaining data cards, start next program.

A. GANT'S METHOD

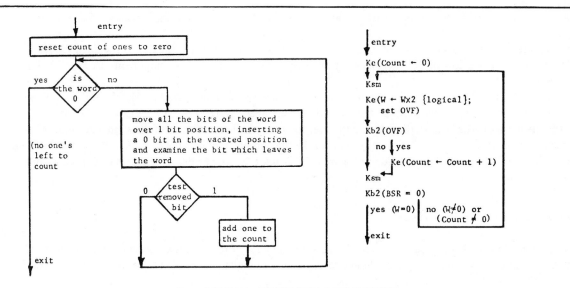

B. DIGITAL EQUIPMENT CORPORATION

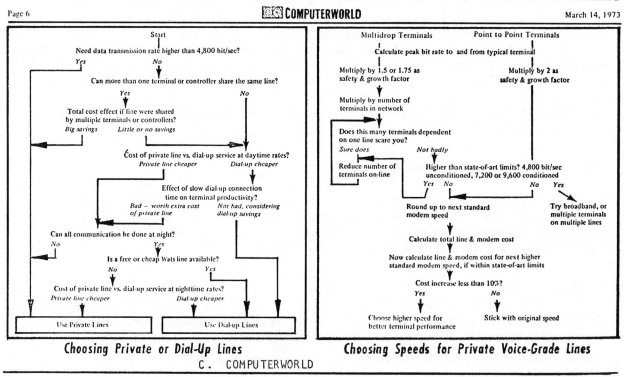

Choosing Private or Dial-Up Lines   *Choosing Speeds for Private Voice-Grade Lines*

C. COMPUTERWORLD

FIGURE 8.2 *Flowchart Outlining*

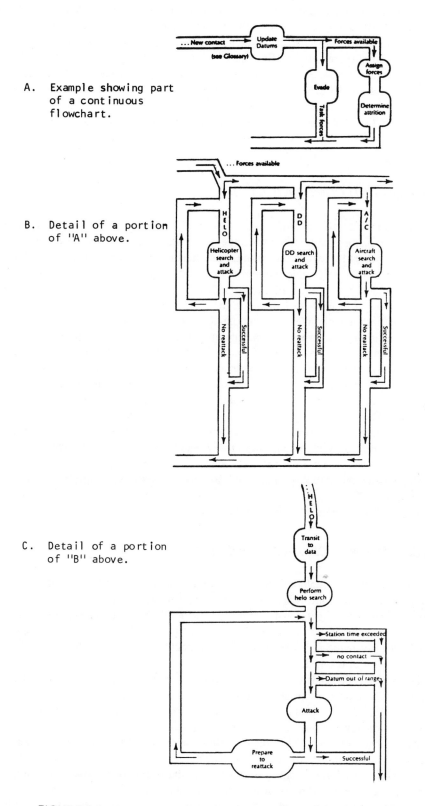

A. Example showing part of a continuous flowchart.

B. Detail of a portion of "A" above.

C. Detail of a portion of "B" above.

FIGURE 8.3 Gieszl's Continuous Flowcharting

conventions and some, but not all, the ANSI standard symbols. They give the nonappearance of flowcharting in their respective technology. Meanwhile, there are numerous flow diagrams that are flowcharts and still others that are called flowcharts, but by no stretch of the imagination could they qualify as such.

The preponderance of flowcharting in the noncomputing field is easy to understand. In most cases where logic, reasoning, and planning are required, flowcharting is a means of organizing one's thoughts on paper and visually examining the total solution.[19] Courses on problem solving are dependent on flowcharting to organize and present a solution to posed problems. Since computer program listings are difficult to analyze, most instructors in programming courses suggest or require that a flowchart be submitted by students who are solving programming or coding problems. Obviously, as one obtains confidence and skill in flowcharting, he will apply the technique more readily to organize his thoughts and go through a series of penetrating questions and answers before he makes a final decision.

EDUCATION AND COMMUNICATION

Flowcharting in the educational field frequently can outperform written text as a means for conveying ideas. There is no speed reading method known to the writer where a page of text material can be read and correlated as easily as a flowchart for logic presentation. Flowcharting has become a language for communication and expression of ideas that is dynamic and in many cases self-explanatory. In many textbooks the reader will find lengthy pages of description summarized with a flowchart. A typical flowchart in math on a fast Fourier transform is shown in Figure 8.4. Sometimes, the word description is reduced and the writer is dependent largely on the flowchart as the major means of communication to his reader.

Flowcharting is taught in every freshman computer science curriculum either as a separate course or as part of some programming course. The flowchart course content is the vehicle used by the academic institution to instruct the student in how to organize his thoughts in a logical manner and express them in flowcharts. The student is encouraged to apply the same principles for problem solving of any of his task assignments including areas in the computer field. The course content is a way of introducing the student to the complex field of computer algorithms and modelling.[19]

As in the evolutionary process of high school math, flowcharting will be repeated. First year college (differential and integral) calculus is commonplace in many high schools, and former high school subjects are introduced at the junior high level. The present first year college computer science course will become a regular high school course. The present high school computer and flowchart courses will become regular courses at the junior high level.[30] Students will be flowcharting their homework solutions rather than outlining them. The use of flowcharts in text books will undoubtedly continue to increase.

BUSINESS, COMMERCE, AND INDUSTRY

Any business firm which installs a digital data processor, a management information system, a process control system, or a digital communication network proves vulnerable to still further penetration by computer technology applications. Once the computer installation has satisfied the initial purchase requirement, more and more uses are found to maximize the owner's return on his investment. If the work load exceeds the computer facility capacity, the work load is not reduced. Instead the computer facility is upgraded to handle the large load and reserve capacity is installed or planned.

Standardization is demanded to prevent chaos. Business wants standards for freedom of choice although some manufacturers, because of proprietary skills and products, prefer a captive market. Standards simplify interchangeability and encourage investments in the computer industry. Regardless of the degree of standardization attained, everyone from top down in business, commerce, and industry is exposed to data processing in some fashion. Anyone intending to enter these fields would do well to be knowledgeable in computer technology.

A quick way of obtaining information on any subject is by means of flowcharts. A typical case is in the making. Today every librarian's formal training includes computer technology. Initially the subject was briefly mentioned and introduced and some text books may have included an appendix on flowcharting. The gist of the librarian curriculum was that a computer did exist and could make a contribution in their field.* Now, the flowchart symbols and their use have moved from the rear (appendix) to the front of most text books. Flowcharting occupies a

*Sally Swenson, "Flowchart on Library Search Techniques," *Special Libraries*, Vol. 56, No. 4 (1965), pp. 239-242.

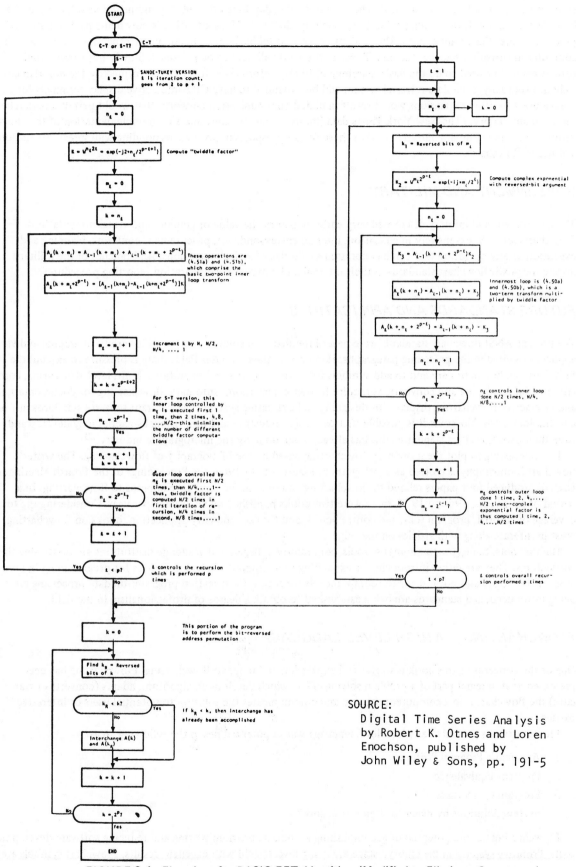

FIGURE 8.4   Flowchart for BASIC FFT Algorithm (Modified to Eliminate Connectors)

complete chapter or large portion of it. The author, having developed the subject and introduced it early in the book, is free to use flowcharting profusely throughout the book. More attention is now being paid to automatic processes rather than manual ones. The student librarian is taking courses in computer programming along with computer information science students. Those in the field will have to be retrained. Slowly but surely, mini-computers will be used to do the basic searching of library information formerly performed by the librarian custodian. Eventually, such data retrieval system will be expanded to have CRT interactive display terminals for library users. Further expansions would result in digital communication networks that would permit a researcher access to any library. (The New York Times data library system of subscribers is an implementation of the above future public and company libraries.) All of these future prospects imply a corresponding widespread generation and use of flowcharts.

GOVERNMENT AND MILITARY

The government and military discovered very early in history the value of graphics and visual presentations. Flowcharts are used extensively in education, internal correspondence, procurement contracts, planning and execution of any operation that can be expressed as a series of logical operations. All branches of the military invoke the ANSI flowchart standards in their internal and external documentation reporting procedures.

FUTURE STANDARDS AND APPLICATIONS

The present ANSI flowchart standards were never intended to accommodate the wide variety of graphic and visual applications cited in the preceding paragraphs. However, in the computer field, there is pressure to expand the flowchart symbol dictionary and to add symbols for more processes and operations. Multi-level structuring, correlation coding, indexing and conventions require further explanation. Automatic flowcharting for documentation and interactive flowcharting displays for designing and evaluating systems may warrant separate standards just on a technology basis. Moreover it is possible that particular industries and professional societies may develop and issue their own flowchart standards; this has already been done by the micrographic industry.[20]

It is interesting to plot the growth of flowcharting based on the bibliography of flowcharting. The writer's flowchart bibliography is plotted as a histogram in Figure 8.5. In 1965, the Proceedings of the Fourth Meeting of UAIDE produced four papers related to flowcharting and over all the years gave impetus to flowcharting. In a period of four years, 1968 to 1972, there were two books published on an average per year. Considering the total coverage of tutorial, program text, and course books, and for each industry application, books on flowcharting must proliferate along with articles on the subject.

High schools, trade (programming) schools, community colleges, and academic institutions are reassessing their curriculums. They are either introducing or expanding their computer sciences courses for programming in numerous technologies and not just for the computer science major. Certification programs on data processing are being introduced, and standards are being established to obtain a degree of professionalism in the field.

FLOWCHARTING – A HIGH LEVEL LANGUAGE

One of the concerns of this book is to give full expression and range to flowcharting. Flowcharting has been presented as an integral part of a problem solving effort which develops an algorithm, adapts (converts or translates) the flowchart into a computer program, and communicates the solution to potential users or interested parties.

Flowcharting is presented as a high level language whose potential lies in the following areas:

1. Automatic Program Writing
2. Program Equivalence
3. Program Correctness
4. System Solutions by means of Items 1, 2, and 3

Flowcharting has the potential of accomplishing automatic program writing and reducing software development costs. Primitive systems in the embryo stage are being investigated with excellent results. Flowchart symbols are

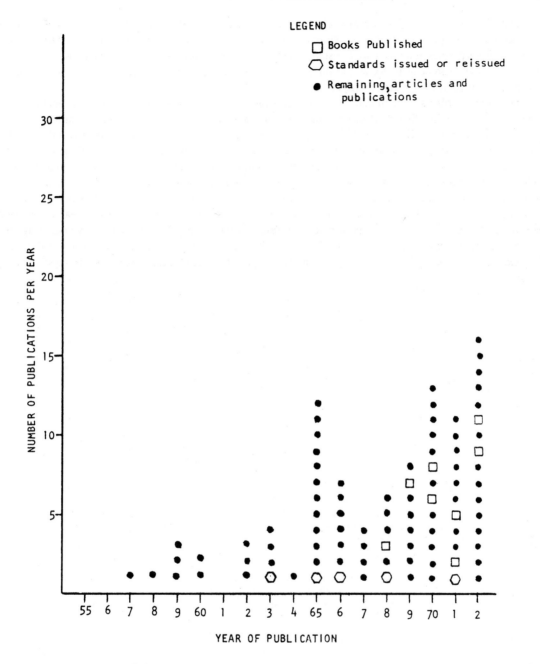

FIGURE 8.5   Flowchart Bibliography Histogram

being defined by rules of formalization, and current work is aimed at compiling from the flowchart representation. The hierarchial structure permits the flexibility of random up and down hierarchy with interactive feedback from lower levels to higher levels (see Chapter 3 for multi-levels, Chapter 4 for multi-tasking, and Chapter 5 for top down and bottom up compiling) using three basic programming operations (see Figure 5.2). The bidirectional mapping (see Chapter 1) when applied to high level languages (see Chapter 6) for DO, GO TO, and IF statements permits the configuration or production representation of many logical processes.

Program equivalence and similarities are mostly being done manually; visually by the programmer. Chapter 6 translated four high level language statements of DO, GO TO, and IF statements into flowchart form. The manual examination revealed equivalences and similarities (see work problem on digital elements). A new high level language can be flowcharted and compared with a known high-level language for equivalence to assess the new language capabilities. When this process is extended to include any flowchart or schema, a general theory of flowcharts can be formalized and the comparing process can be automated for pattern equivalence and similarities.

Algorithm analysis together with program correctness and program equivalence are interrelated and interdependent. The techniques and formalization rules of axiom proofs with flowcharts are used to obtain flowchart equivalence when one flowchart serves as a reference (or standard) for comparison with another flowchart. Recursively applied, a series of flowcharts can be analyzed and evaluated. When program equivalence cannot be obtained, the algorithm proofs and program correctness are employed.

Solving system problems is the integration and utilization of the first three items. A CRT terminal display in conjunction with an on-line flowchart programming system capability provides the ultimate in automatic program writing and system solution. Here, not only can software be created, it can be tested, debugged, and peaked if necessary. Parametric range values, processing time (e.g., add time, fetch time, I/O processor time, etc.), and percentage selection values for each of the multiple symbol outputs (e.g., branch table, decision symbol, and switch operations) together are computable to ascertain the degree of system stability. Inputs to the system can be progressively added and/or parametric values altered to disclose system saturation, time limitation, or deadlock situations. Alternate methods can be substituted (look up table instead of a computation) to peak up the system performance. The turn around time in an interactive terminal display using flowcharting for system design is quite short compared with current equivalent manual methods. The dividends are so great that industry and government funding are becoming available for research and building exploratory automatic program writing systems using flowcharts.

Appendix A

FLOWCHART PROGRAMMING SYSTEMS (INTERACTIVE)

Interactive flowchart programming systems are viewed from a dialogue vantage point for generationg programs from flowcharts. A number of experimental flowchart programming systems (FPS) are present and some are in the design stage.[4,21-23]

The interactive FPS is of interest, because it assists the programmer (user/operator) in four major areas of programming:

1. Designing and developing program units
2. Assembling program units into larger units
3. Checking, testing and debugging the program units throughout each stage of design and development
4. Documentation of the complete design process and program

The areas of interest are flowchart techniques, flowchart symbols, conventions, and notations of two instructive flowchart programming systems.[4,22] The flowchart symbols for the GRAIL language and the Brown University FPS are shown in Figure 8.6 and 8.7, respectively. The FPS permits construction, editing, interpretive execution, compilation, debugging, documentation, and execution of computer programs by flowcharts. The flowchart language encompasses the constraints of the system configuration, computer software, the unpredictable actions of the user, the display devices, geometry of the flowchart symbols, and flowchart hierarchy structure.

In the GRAIL system, each program level has to be defined. The top level and their interrelationships constitute a conceptual plane. The next level of detail for a particular notion constitutes another conceptual plane and so on, until the lowest level of detail has been explicitly defined by an appropriate program statement or flowchart symbol. A collection of planes defines a computer process. The hierarchy is an amalgamation of program units, display form (frame pictures), flowchart elements, and computer processes (parent-daughter context). The GRAIL system is an early experimental model in the FPS and is often cited.

The flowchart programming system at Brown University shall be called by the letter string FPS. FPS, like GRAIL, developed a unique set of flowchart symbols (15) that are shown in Figure 8.7. Briefly, this system allows the user to specify his algorithm as a flowchart drawn on a display (IBM 2250 display unit) to be translated into conventional source text (e.g., PL/1, Assembly Language, FORTRAN) by a flowchart compiler.

The user interacts with an editing system to build a representation of his program, which he can use on-line to design, determine and remove errors, and document his work. Initially, either the user may conceive and create his program wholly on-line using the system or he may already have a rough version of a flowchart when he sits down, so that he will transcribe and improve his program during his input session. When his program or part of his program is completed to his satisfaction, the user requests that the flowchart compiler process his program. When the compiler is done, the user can call the generated target code to make changes and additions and recompile his program on line. After producing target language code from his program by using the compiler, the user passes this code to a standard language processor. It should be noted here, the target language is the flowchart compiler output that is the source language for a conventional compiler. There is a one-to-one mapping between the display screen (flowchart) and target language code. At this point, if any corrections are to be made, the user has the choice of a more efficient program or the corresponding fidelity between the flowchart (developed on the screen) and the coding. If the user proposes to edit, correct, or improve the generated code at this stage, it will not reflect the flowchart source. A preferable solution would be to change the flowchart of the program and reprocess it with the flowchart compiler. A similar situation exists in program testing. The programmer should avoid making corrections in the object deck. This guarantees that no discrepancies between the source and object deck can occur and eliminates any serious errors in program production and maintenance.

The final output of the compiler is the generated target language. Figure 8.8 shows the PL/1 and Assembly Language flowcharts and their corresponding code. From here, a punch card and listing can be made. Also, the program can be entered and run. At all times, error tracking is maintained from the compiler to the language processor output via the target code. Any encountered errors can be viewed online and edited so that the user can delete them as he makes individual corrections.

Using the FPS the user can proceed from a vague system concept through more and more detail until an explicit representation of the operation in flowchart form has been developed. As each program element is

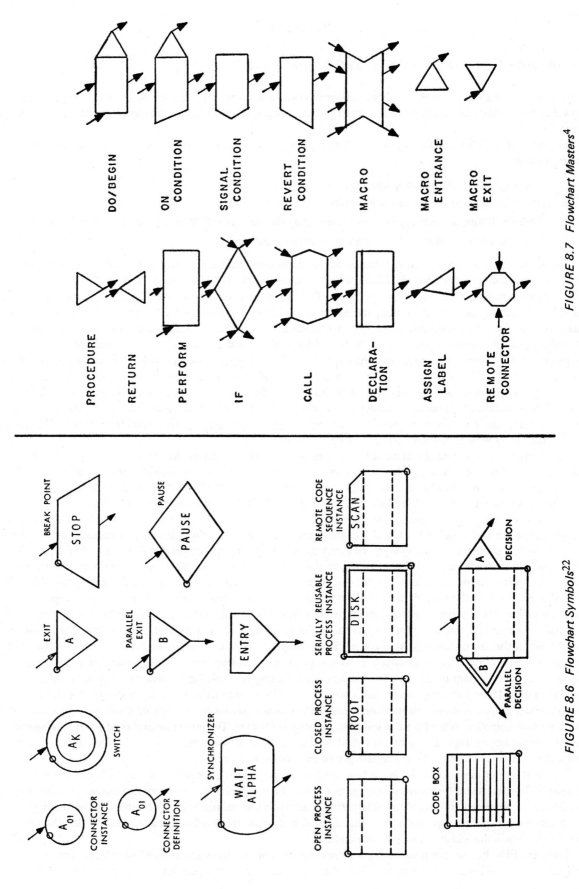

FIGURE 8.7   Flowchart Masters[4]

FIGURE 8.6   Flowchart Symbols[22]

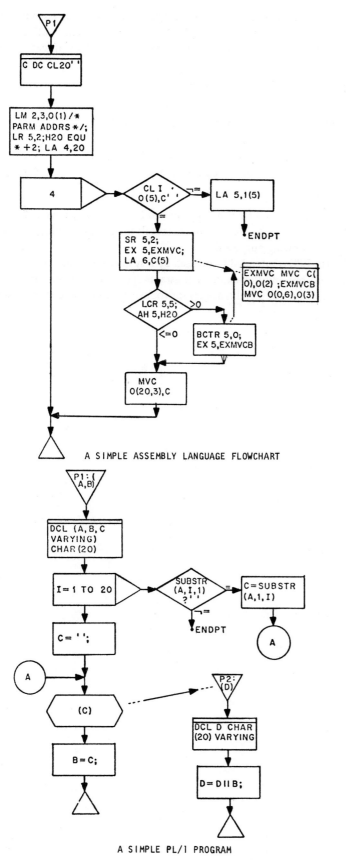

A SIMPLE ASSEMBLY LANGUAGE FLOWCHART

CODE GENERATED

```
P1          ENTER
*  GET PARM ADDRESS
            LM     2,3,0(1)
            LR     5,2
            LA     4,20
H20         EQU    *-2
            LTR    4,4
            BP     A2700380
            B      A27004B0
A2700440    BCT    4,*+8
            B      A27004B0
A2700380    CLI    0(5),C' '
            BC     8,A27005C0
            LA     5,1(5)
            B      A2700440
A27005C0    SR     5,2
            EX     5,EXMVC
            LA     6,C(5)
            LCR    5,5
            AH     5,H20
            BC     12,A2700220
            BCTR   5,0
            EX     5,EXMVCB
A2700220    MVC    0(20,3),C
A27004B0    L      13,4(13)
            RETURN (14,12),T,RC=0
C           DC     CL20' '
EXMVC       MVC    C(0),0(2)
EXMVCB      MVC    0(0,6),0(3)
            END
```

A SIMPLE PL/1 PROGRAM

CODE GENERATED

```
P1: PROCEDURE (A,B);
 DCL (A,B,C VARYING) CHAR(20);
 DO I=1 TO 20;
 IF SUBSTR (A,I,1)¬=' ' THEN GO TO A27003C0;
 C=SUBSTR(A,1,I);
 GO TO A2700480;
A27003C0: END;
 C=' ';
A2700480: CALL P2 (C);
 B=C;
 RETURN;
P2: PROCEDURE (D);
 DCL D CHAR(20) VARYING;
 D=D||B;
 RETURN;
 END P2;
 END P1;
```

FIGURE 8.8[4]

completed, it is stored and recalled in the correct sequence when the segment, thread, or function is assembled for final examination prior to running the computer program. During the flowcharting process, existing routines and algorithms on file (in the library) can be called and inserted in the proper sequence. This flowchart process can be summarized in the vernacular of programming, compiling and flowchart drawing.

		PROGRAM/COMPILE	*FLOWCHART*
1.	TOP DOWN	Start at the top and work down to the lowest level of detail.	Start at the input requirements and work toward the output.
2.	BOTTOM UP	Start at the lowest detail and work up toward the top.	Start at the output requirements and work toward the front end (input requirements).
3.	HYBRID	Combination of Items 1 and 2.	

As the bottom up construction proceeds, the program (and flowchart) is limited by the display area and the program complexity increases. The user wants to be able to manage the whole (total) program by maintaining as much detail as possible on a limited display area for viewing purposes.

The FPS is a more flexible system than GRAIL. FPS is intended to be used by a system programmer. The system includes various standard flowcharting notations and flowcharting constants specially designed for such a user. This may be the reason for using a single symbol (equilateral triangle) for four operations: procedure; return; macro entrance; and macro exit.

Appendix B

ANALYSIS AND SYNTHESIS OF ALGORITHMS BY FLOWCHARTS

Probably, flowchart utilization and application will find its greatest acceptance in areas not associated with software documentation. Apparently, flowcharts are now fashionable in analyzing algorithms, decomposing (reducing) detailed programs, providing program equivalence, formulating and illustrating structured programs, and graphically presenting details that are too difficult to explain by equations and test. Flowcharts can combine or partition individually control flow and information flow. Series and parallelism are also available. Problem areas can be isolated, examined and resolved with a minimum of flowchart symbols. In turn, segments can be formed of primitive elements for any length until complete programs are configured. The reverse is also true; complete programs can be reduced to a basic concept. The result is the application of flowcharts as the prime means of formulation, analysis, evaluation and recommendation.

ALGORITHM ANALYSIS AND DECOMPOSITION

The term "algorithm" is used in many different ways. Sometimes it is used as a process in abstract without reference to a particular computer. Generally speaking, an algorithm is any particular procedure used to solve a problem. Hence, a particular sequence of instructions or its equivalent expressed as an equation or function is considered an algorithmic notation. Theorems are defined in algorithmic notations independently of expression in a given programming language. Theorems proving program correctness, verification of programs, and the theory of program schemas exhibit some relationship among algorithms and flowcharts. The flowcharts used obey finite directed graph theory. The primary concern is with the finite expression of algorithms by charts that indicate a flow of control/data from one command (assertion) to the next command (assertion). The typical flowchart program schema requires five primitives[24] as defined in Figure 5.1 and repeated here as Figure 8.9.

Floyd[24] speaks of five statement types while Bruno-Steiglitz[25] express their algorithms from command to command in flowchart form using four symbols. Symbol 3 is eliminated by the Bruno-Steiglitz team by combining it with symbols 1 and 2 as follows:

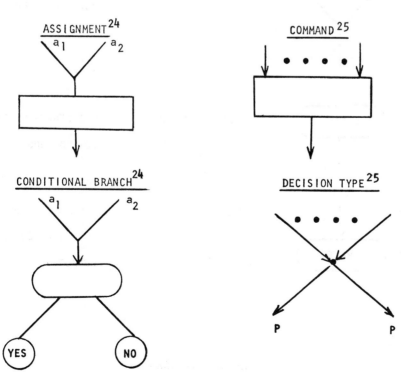

SYMBOL NO.	SYMBOL	IDENTIFICATION	DESCRIPTION

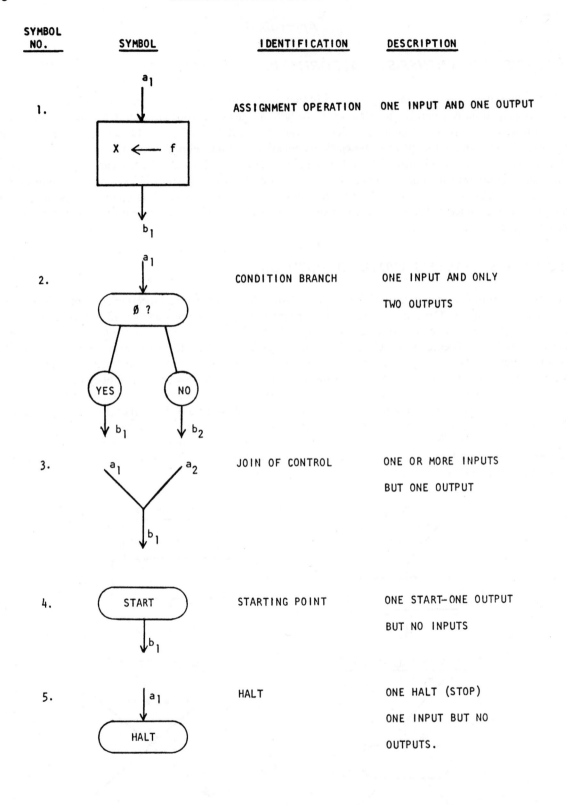

FIGURE 8.9 Five Primitives[24]

PROGRAM EQUIVALENCE

There are a number of definitions for program equivalence. Ershov[26] is rather thorough in defining equivalence, whereas Elspas et al[27] simplify the kinds of equivalences. The approach taken here is a flowchart implementation based on stated axioms and proven theorems in the literature.[25-27]

In obtaining a reference model for comparison or in comparing an unknown with a reference for identification, the following "blackbox" procedure is taken. The only accessibility that is available is the interface of the item under test with the outside world; entry and exit. Assuming the unknown is exhaustively tested (every possible internal state and sequence), the only method of identification is the outputs as a function of the inputs. If another item is exhaustively tested and the results of the tests are the same, it is inferred that the test item and the reference are identical by the following equation:

$$\text{output} \leftarrow (\text{input}) \; X \; (\text{internal state})$$

Since the measurable quantities of both are alike in every respect, it follows the internal state of both are alike. The supposition of exhaustive testing, a goal and not a fact, does not detract from the rigor that both items are alike. If it cannot be discerned there is difference between two items by testing and one can be substituted for the other for purposes of utilization, they are alike in every respect because the results are alike in every respect. If a test cannot be devised to obtain a difference between the two items or the test procedure is limited by the test facility, then the two items are assumed to be identical until disproven. The latter has happened occasionally, but this does not detract from the following equation:

$$S_R \equiv S_X$$

This relationship has very practical significance, since it may enable us to prove something (e.g., correctness) about one program by showing that this program is equivalent to another one whose correctness (or other measurements of correctness)[26] has been established. The concept of equivalence is significant from a theoretical point of view as well, if only because it involves questions of decidability and partial decidability for various kinds of flowchart schemas and for various kinds of equivalences. Table 8.1 lists various degrees of equivalence defined in terms of inputs and outputs. The material is a selected extraction from Reference 27.*

In Chapter 6, DO type statements were examined for four high level languages, FORTRAN, ALGOL, COBOL, and PL/1. Each flowchart for each language statement has the same symbols and symbol arrangement (or expresses the same transformation) and they are equivalent. Given the appropriate input variables (neglecting the compiler subtleties), the outputs of each are the same. Here, the inputs, outputs, symbols, and flowlines are matched up on a per item basis for each language and the comparison output would be true for each test. This is not surprising since it was expected. Any new language statement capability could be evaluated with the mentioned flowchart schema for comparison and tested on the equivalence criteria. A flowchart can be used to express an algorithm, independent of machine and language. As a matter of fact, this is done in specifying a software requirement. If the flowchart is implemented by each of the four high level languages, they are required to operate on the same input variables and deliver the same results. These four high level language flowchart representations express the same transformation, whatever the meaning of the flowchart symbols contained in it. Were it not for the syntax and semantics of the particular language, the blank outlines in the flowchart could be overlayed for each language and the chances are they would form one transparency. The logic (sequence) would be the same since they implemented the same flowchart (specification). This meets the strong equivalence

$$S \equiv S'$$

both are defined and Val (S, I, X) = Val (S', I, X). Here the interpretations of each specification and implementation, are the same.

Let's examine a weak equivalence and upgrade it to a strong equivalence. Two programs are equal when the inputs are equal and the outputs are equal. The two flowchart schemas S and S' are said to be weakly equivalent (written $S \simeq S'$) if they are similar, but not equal. If one is upgraded to be general purpose and performs all the requirements of the special purpose, the equivalence becomes strong equivalance. Upon testing the general purpose

*Kinds of Equivalence, pp. 106-108.

Table 8.1. Kinds of Equivalences*

DEFINITION: Two flowchart schemas are said to be *compatible* if they have the same number of input variables and the same number of output variables. (Note that they are allowed to have entirely different sets of internal program variables.)

DEFINITION: Two compatible schemas S and S′ are said to be *strongly equivalent* (written S ≡ S′) if under every interpretation I and input assignment x, *either:*

1. *both* Val (S, I, x) and Val (S′, I, x) are undefined *or*
2. *both* are defined and Val (S, I, x) = Val (S′, I, x)

DEFINITION: Two compatible flowchart schemas S and S′ are said to be *weakly equivalent* (written S ≃ S) if under every interpretation I and input assignment x, *either:*

1. Val (S, I, x) or Val (S′, I, x) is undefined *or*
2. both are defined and Val (S, I, x) = Val (S′, I, x)

DEFINITION: Two programs P and P′ are said to be *termination equivalent* if, for the same input values x, both P and P′ terminate *and* yield the same result (i.e., P(x) = P′(x)).

DEFINITION: Two compatible schemas S and S′ are related by S > S′ (i.e., S is said to be an *extension* of S′) if: whenever Val (S′, I, x) is defined so is Val (S, I, x) and, moreover, Val (S, I, x) = Val (S′, I, x).

DEFINITION: A relation R between schemas is said to be *reasonable* if for every pair of compatible schemas S, S′, both of the following are true:

1. if S ≡ S′ then S R S′ *and*
2. if S R S′ then S ≃ S′

program at the special purpose requirements, the equivalence criteria is met; S ≡ S′. The three program structures are used to reduce either one or both to obtain equivalence (F ≡ F′). A top down parsing and bottom up parsing were shown in Figure 5.3. A number of reductions (bottom up) or productions (top down) on both flowcharts are done with the aim of obtaining F ≡ F′. The one input and one output concept is retained while insuring the same interpretations and input values (by concatenation).

If it is agreed that flowcharts can be parsed in either direction, then it follows that logic processing can be generated in the same manner that programs are generated from formal definitions and notations. The program structures with construction rules will permit the generation of programs by flowchart means (e.g., interactive displays). There is a need for a formal definition of a high level flowchart language having the BNF formal metalanguage for phrase structure grammars.[25]

For the lack of a better definition, a program may be said to be correct if its execution terminates and yields a desired final result (program ran to completion). This is a vague definition since program correctness did not include performance and reliability. In some cases, these terms are not measurable and therein lies the problem. With respect to program correctness, there is no foolproof means for verifying the correctness of a program or a program segment. Since we can never completely prove that we have found the last error, we can only say, after "all discoverable" errors have been corrected, the test disclosed errors and not the absence of errors (the program may not be error free).

Most program proving methods and program correction techniques employ mathematical logic, propositional calculus, and predicate calculus. Reference 27 is an excellent survey on the subject of program correctness, while Reference 28 is an assessment of these techniques.

PROGRAM CORRECTNESS BY AUTOMATIC MEANS

After a flowchart has been drawn, it is examined for correctness in two areas:

1. Logically correct
2. Solution to the problem or meets specifications

*See Reference 27.

The second item is one of interpretation and not applicable here, but logical program correctness is. The logic correctness is associated with data flow and control flow when the programmer or systems analysis modifies and corrects his flowchart. There are certain criteria he applies to the flowchart to test it to discover any errors for this terminology. A few of these rules that he applies to evaluating his flowchart for correctness can be stored and applied to check the flowchart (and, therefore, the program) for correctness. There are some specific logical sequences that have been proven correct and can be stored as a string (a series of numbers). The flowchart is automatically read as a string and checked for correctness. There are a number of standardized and correct processing techniques and also a number of incorrect or faulty procedures. They are stored in a look up table and used to check out the logic of a program. Whatever data sheets and procedure rules the programmer used to evaluate his flowchart can be mechanized and automatized. In compiling, syntax errors, undefined constants and variables, and simple logic errors are caught and brought to the attention of the programmer. There are always some system constraints and ground rules. Instead of using reference tables, hand calculations and criteria checking rules by manual means and subject to human error, they can be automatized and applied error free in most cases.

REFERENCES

1. Boehm, Barry W. "Software and its impact: A Quantitative Assessment." *Datamation,* May 1973, pp. 48-59.

2. Maycock, P.; Tory, A. C.; and Wintle, R. N. "Automatic Coding From A Flow Diagram," Decus Europe 8th Seminar Proceeding, Strasbourg, France, 20-22 Sept. 1972, pp. 381-3.

3. Williams, T. C. and Bebb, Joan, "An On-Line System For Hand-Printed Input," Rand Report TM-4786, August 30, 1971, 39 pages.

4. Michener, J. C. "The Flow Chart Programming System," Ph.D. Thesis, Brown University, June 1970, 229 pages.

5. "Talking Down a Program." *Business Automation,* May 1972, pp. 26-27.

6. Scott, A. E. "Automatic Preparation of Flow Chart Listing." *Jr. ACM,* January 1958, pp. 57-66.

7. Haibt, Louis M. "A Program to Draw Multilevel Flow Charts," Proc. Western Joint Comp. Conf. 1959, pp. 131-137.

8. Knuth, Donald E. "A Computer-Drawn Flowcharts." *Comm. ACM,* September 1963, pp. 555-563.

9. The following papers were presented at the Proceeding of the Fourth Meeting of UAIDE, October 1965.

   Goetz, Martin A. "Recent Developments in Automated Program Documentation." pp. XIX 1-15.

   Hain, G., and Hain, K. "A General Purpose Automatic Flowcharter," pp. IV i-ii, IV 1-12.

   Roberts, G. V. "Automated Diagram Documentation (ADD)." pp. XXIII 1-20.

   Sallbach, C. P., and Sapovchak, B. J. "The Flowchart Program," pp. XXII 1-7.

10. Auerbach Computer Technology Reports "Definitional Reports-Documentation Aids." May 1972, pp. 1-13; "Comparison Charts," March 1972, p. 106.

11. Falor, Ken. "Survey of Program Packages-Programming Aids." *Modern Data Magazine,* March 1970, Table 4 Documentation and Flwochart-Preliminary Programs, p. 68.

12. Chapin, Ned. "Flowchart Packages." *Data Management,* October 1970, pp. 16-23.

    "Running Time Analysis for Flowcharts," *Software Age,* Feb/Mar 1971, pp. 13-15, 30.

    "Perspective for Flowcharting Packages." *Computer and Automation,* March 1971, pp. 16-19, 26.

    "Flowcharts." Pub. Auerbach, 1971, Chapter 6, pp. 137-145.

    "Flowchart Packages and ANSI Standard." *Datamation,* September 1972, pp. 48-53.

13. Peterson, Norman D. "An 'Atlas' for the Visual Comparison of COBOL Flowcharting Software." IEEE Repository Paper #R71-129, May 1971, 94 pages.

14. Bell, C. Gordon; Grason, John; and Newell, Allen. *Design Computers and Digital Systems.* Digital Press, 1972.

15. Hardy, Robert M "A Flow Chart Hardware Correlation." *Digital Design,* June 1973, pp. 46-47.

16. Bjorner, Dines. "Flowchart Machines." IBM Research RJ 685(#13346), April 7, 1970, 46 pages.

17. Gant, W. T. "Flow-outlining A Substitute For Flow Charting." *Comm. ACM,* November 1959, p. 17.

18. Gieszl, Louis R. "Continuous Flow Charts." *Simulation,* June 1970, pp. 281-289.

19. Honeycutt, T. L. "Teaching the Art of Flow Charting in Computerized Problem Solving." *J. Educ. Data Processing,* Vol. 9, No. 3, 1972, pp. 19-28.

20. National Microfilm Association. "Flowchart Symbols and their Usage in Micrographics." NMA MS4-1972.

21. Sutherland, W. R. "On-Line Graphical Specification of Computer Procedures." MIT Lincoln Laboratory Technical Report 405, ESP-TR-66-211, 23 May 1966.

22. Ellis, T. O.; Heafner, J. R.; and Sibley, W. L. "The GRAIL Language and Operations." The Rand Corporation, RM-6001, ARPA, (ARPA Order No. 189-1), September 1969, 23 pages.

23. Bernstein, M. I. "The Design For An Interactive Flowchart Programming System." Proc. of the 3rd Hawaii International Conference on System Science, Honolulu, Hawaii, January 1970, pp. 894-897.

24. Floyd, Robert W. "Assigning Meanings to Programs." Proc. of Symposium in Applied Mathematics of the American Mathematical Society, April 5-7, 1966, pp. 19-32.

25. Bruno, J. and Steilglitz, K. "The Expression of Algorithms by Charts." *Jr. of the Assoc. of Comp. Mach.*, July 1972, pp. 517-523.

26. Ershov, A. P., "Theory of Program Schemata." *Information Processing 71*. North-Holland Publishing Company, 1972, pp. 28-45.

27. Elspas, Bernard; Levitt, Karl N.; Waldinger, Richard J.; and Waksman, Abraham. "An Assessment of Techniques for Proving Program Correctness." *ACM Computer Surveys,* June 1972, pp. 97-147.

28. Prokop, J. S. "On Proving the Correctness of Computer Programs." *Program Test Methods,* edited by Willaim G. Hetzel, Prentice-Hall, 1973, pp. 29-37.

29. Rice, John K. and Rice, John R. "Introduction to Computer Science." Holt, Rinehart, and Winston, Inc., 1969, p. 65.

30. McQuigg, James D.; and Harness, Alta M. "Flowcharting" Houghton Miffin Co., 1970, 43 pages.

31. Goldberg, Patricia; "Automatic Programming," in Programming Methodology, Vol. 23, Lecture Notes in Computer Science, Edited by Clemens E. Hackl, Springer-Verlag, 1975, pp. 347-361.

32. Reifer, Donald J., "Automated Aids For Reliable Software." Proc. International Conference on Reliable Software, April 21-23, 1975, Los Angeles, Calif., 136.

1   PROBLEMS

1. Code the following flowchart in any language (FORTRAN, COBOL, PL/1, ALGOL, BAL, etc.). The read-in and print operations are optional, Flowchart the program statements and compare with the flowchart below. If necessary decompose (or reduce) the flowchart listing and prove program equivalence if possible.

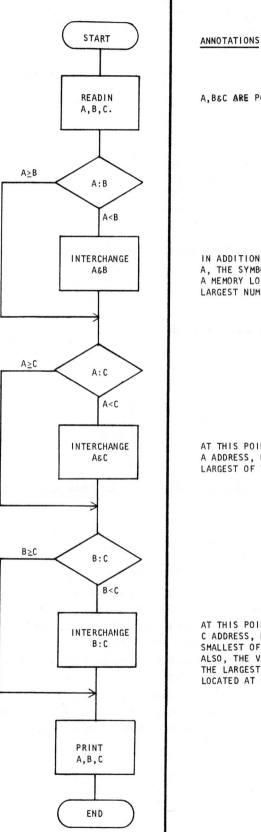

ANNOTATIONS

A,B&C ARE POSITIVE NUMERICS

IN ADDITION TO THE NUMERIC
A, THE SYMBOL A NOW REFERS TO
A MEMORY LOCATION FOR THE
LARGEST NUMBER

AT THIS POINT, THE SYMBOLIC
A ADDRESS, NOW CONTAINS THE
LARGEST OF THE 3 INPUT NUMBERS

AT THIS POINT, THE SYMBOLIC
C ADDRESS, NOW CONTAINS THE
SMALLEST OF THE 3 INPUT NUMBERS.
ALSO, THE VALUE THAT IS NEITHER
THE LARGEST NOR THE SMALLEST IS
LOCATED AT THE B ADDRESS.

2. Extend Problem 1 to contain a list of N instead of three positive numbers. Modify the flowchart in Problem 1 to accommodate N items. Examine the sort flowchart carefully. Now reduce the sort flowchart to be equal to the one presented in Problem 1. (Hint: Use the three basic programming operation modules).

ANSWER

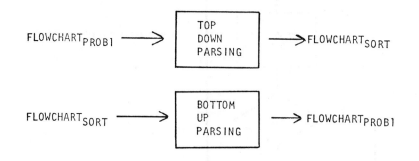

3. Programming logic can be represented by three basic elements:

   1. Concatenation/Sequence
   2. Selection/Branch
   3. Repeat/Loop

Digital logic can be represented by three basic elements:

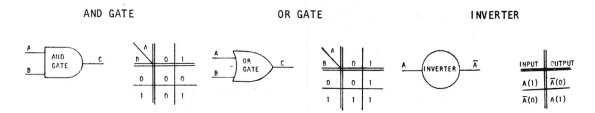

Flowchart the truth tables and examine the logic relationship of the digital elements: Are they equivalent, similar or not related?

Code the flowcharts using the instruction code set below.

INSTRUCTION CODE SET

OP CODE	OPERATION	DETAILED DESCRIPTION
CLA	Clear and Add	Take a word from storage specified by the address portion of CLA and place it in the accumulator.
STO	Store	Take a word from the accumulator and place it in storage location specified by the address portion of STO.
TRA	Transfer	Address part specifies the location of the next instruction to be executed — unconditional transfer.
TRZ	Transfer on Zero	If test is true (accumulator contains a zero), the next instruction is taken from the location specified by the address portion of TRZ, otherwise execute the next instruction — conditional transfer.

ANSWER

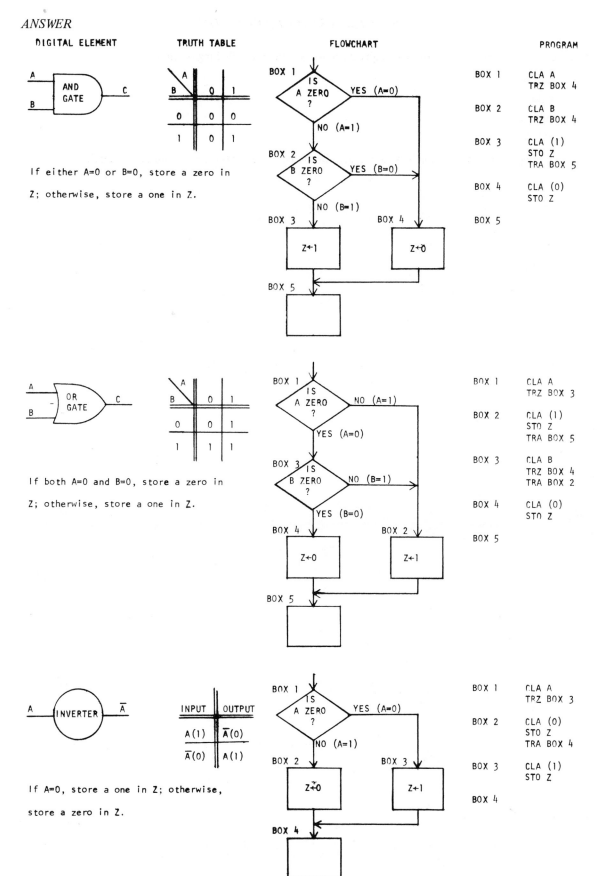

| DIGITAL ELEMENT | TRUTH TABLE | FLOWCHART | PROGRAM |

If either A=0 or B=0, store a zero in Z; otherwise, store a one in Z.

If both A=0 and B=0, store a zero in Z; otherwise, store a one in Z.

If A=0, store a one in Z; otherwise, store a zero in Z.

FLOWCHART BIBLIOGRAPHY

An extensive bibliography on flowcharting is given here having the following outline

1. Author(s)
2. Title
3. Publication
4. References Cited

The bibliography is not complete by any means but is available for the serious minded researcher.

FLOWCHART BIBLIOGRAPHY

1. Abrams, Marshall D.

    The operation and use of MADFLOW.

    Dept. Eng. Univ. of Maryland, 1968.

    References:

2. Abrams, Marshall D.

    A comparative sampling of the systems for
    producing computer-drawn flowcharts.

    Proc. ACM 23rd National Conference, 1968,
    pp. 743-750

    References: 10

3. Addyman, A. M., and R. Lane

    Software production using FLOCODER

    Conference on Software Engineering for Tele-
    communication and Switching Systems, Colchester,
    Essex, England, April 2-5, 1973, pp. 253-261.

    References: 3

4. American Standards Institute

    Proposed American standard flowchart symbols for
    information processing.

    Comm. ACM Vol. 6., No. 10., October 1963,
    pp. 601-604.

    References: None

5. Anderson, H. E.

    Automated plotting of flow-charts on a small
    computer.

    Comm. ACM Vol. 8., No. 1., January 1965, pp. 38-39.

    References: 3

6. ANSI

    Flowchart symbols and their usage in information
    processing.

    ANSI X3.5 Approved September 1, 1970, copyright 1971,
    17 pp.

    References: None

7. ASA

    ASA standards for flowchart symbols.

    Jr. of Data Management Vol. 3, No. 11, November 1965,
    pp. 18-22.

    References: None

8. Bernstein, M. I.

    The design for an interactive flowchart programming
    system.

    Proc. of the 3rd Hawaii International Conf. on
    System Sciences, Honolulu, Hawaii, Jan. 1970.
    pp. 894-897.

    References: 4

9. Berztiss, A. T., and R. P. Watkins

    Directed graphs and automatic flowcharting.

    Proc. of Fourth Australian Computer Conference,
    Adelaide, South Australia, 1969, pp. 495-499.

    References: 20

10. Bjorner, Dines

     Folded syntax-and recursive flowchart machines,
     a summary.

     Proc. of the 3rd Hawaii International Conf. on
     System Sciences, Jan. 1970, pp. 451-453.

     References: 3

11. Bjorner, Dines

     Flowchart machines

     IBM Research RJ 685 (#13346), April 7, 1970, 46 pp.

     References: 43

12. Bohl, Marilyn

     FLOWCHARTING TECHNIQUES

     Book Pub. by Science Research Associates Inc.,
     1971, 195 pp.

     References: None

13. Bruno, John, and Kenneth Stelglitz

     The expression of algorithms by [flow] charts

     Courant Computer Science Symposium 4- Algorithm
     Specification - Edited by Randall Rustin,
     Prentice-Hall, 1972, pp. 97-115.

     References: 3

14. Bruno, John, and Kenneth Steiglitz

     The expression of algorithms by [flow] charts

     Jr. of ACM, Vol. 19, No. 3, July 1972, pp. 517-525.

     References: 3

15. Butler, D. D., and O. T. Gatto

     Event-chain flow charting in AUTOSTATE: A new
     version

     Rand Corporation, Corporation RM-4729-PR
     (DDC No. AD 622744) October 1965, 70 pp.

     References: 1

16. Chapin, Ned

     Flowchart with the ANSI standard: A tutorial

     ACM Computing Surveys Vol 2, No. 2, June 1970
     pp. 119-146.

     References: 21

17. Chapin, Ned

    Flowchart packages

    Data Management, Vol. 8, No. 10, October 1970.
    pp. 16-23.

    References: 1

18. Chapin, Ned

    Running time analysis for flowcharts

    Software Age, Vol 5, No. 2, Feb/Mar 1971,
    pp. 13-15, 30.

    References: 1

19. Chapin, Ned

    Perspective for flowcharting packages

    Computers and Automation Vol. 20, No. 3,
    March 1971, pp. 16-19, 26.

    References: 2

20. Chapin, Ned

    FLOWCHARTS

    Book Pub. by Auerbach, 1971, 179 pp.

    References: 35

21. Chapin, Ned

    Flowchart packages and the ANSI Standard

    Datamation, Vol 18, No. 9, September 1972.
    pp. 48-53.

    References: 3

22. Chapin, Ned

    New format for flowcharts

    SOFTWARE Practise & Experience
    Vol. 4, No. 4 October-December 1974,
    pp. 341-357.

    References: 12

23. Chaty, G.

    Flow-diagrams and segmentation of a program

    Canadian Computer Conference Session 1972, (French)
    Montreal, Canada, June 1-3, 1972, pp. 3312-3313.

    References: 9

24. Dickerson, Joyce, and B. R. Heap

    A simple program for drawing flow charts

    National Technical Information Service, AD 720 089
    February 1969, 17 pp.

    References: 8

25. Edwards, Perry

    FLOWCHART & FORTRAN IV

    Book Publ. by McGraw-Hill, 1973, 147 pp.

    References: 13

26. Eggert, John L.

    A flowchart technique for computer-aided design
    of small computers

    Computer Design, Vol. 12, No. 11, November 1973,
    pp. 73-81.

    References: None

27. Elliot, Ronald E.

    PROBLEM SOLVING AND FLOWCHARTING

    Book Pub. by Reston Pub. Co. Inc., 1972, 98 pp.

    References: 5

28. Ellis, T. O., J. F. Heafner, and W. L. Sibley

    The GRAIL language and operations

    Rand Corporation RM-6001-ARPA (Contract No.
    DAHC 15 67 C0111), September 1969, 23 pp.

    References: 11

29. Farina, Mario V.

    FLOWCHARTING

    Book Publ. by Prentice-Hall Inc., 1970, 120 pp.

    References: None

30. Fritz, W. Barkley, et al.

    (SHARE's Ad Hoc Committee on Flow Chart Symbols),
    Proposed standard flow chart symbols

    Comm. ACM Vol. 2, No. 10, October 1959, pp. 17-18

    References: None

31. Gant, W. T.

    Flow outlining-A substitute for flow charting

    Comm. ACM Vol. 2, No. 11, November 1959, p. 17.

    References: None

32. Garfield, Eugene, and Irving H. Sher

    Diagonal display - a new technique for graphic
    representation of complex topological networks

    AF49 (638) - 1547 Air Force of Scientific
    Research, September 30, 1967 - submitted by
    Institute of Scientific Information,
    pp. 111 1-3, V 2, V 11.

    References: 50

33. Gatto, O. T., and E. M. Fairbrother

    AUTOSTATE: Part II - event-chain flow charting

    Rand Corporation, Corporation RM-3320-PR,
    December 1962, 92 pp., Modifications 10 pp.

    References: None

34. Giezl, Louis R.

    Continuous flow charts

    Simulation, Vol. 14, No. 6, June 1970, pp. 281-289.

    References: 5

35. Gleim, George A.

    PROGRAM FLOWCHARTING

    Pub. by Holt, Rinehart and Winston Inc., 1970,
    71 pp.

    References: 6

36. Goetz, Martin A.

    Automated program documentation - AUTOFLOW

    Applied Data Research, Inc., Princeton, N.J.,
    1965.

    References:

37. Goetz, Martin A.

    Recent developments in automated program
    documentation

    Proc. of the Fourth Meeting of UAIDE, New York,
    October 1965, pp. XIX 1-15.

    References: None

38. Gorn, S., et al.

    Standards: A standards working paper - conventions
    for the use of symbols in the preparation of
    flowcharts for informations processing systems

    Comm. ACM Vol. 8, No. 7, July 1965, pp. 439-440.

    References: None

39. Green, M. W., B. Elspas, and K. N. Levitt

    Translation of recursive schemas into label stack
    flow-chart schemas

    Standford Research Institute, Menlo Park,
    California, June 1971, abstract, 5 pp.

    References: 2

40. Grems, Mandalay

    Comparison of flow charts symbols

    Comm. ACM Vol. 3, No. 3, March 1960, pp. 174-175.

    References: None

41. Haibt, Lois M.

    A program to draw multilevel flow charts

    Proc. Western Joint Computer Conference March 3-5
    1959, San Francisco, Calif. pp. 131-137

    References: None

42. Hain, G., and K. Hain

    A general purpose automatic flowcharter

    Proc. of the Fourth Meeting of UAIDE, New York,
    October 1965, pp. IV I-II, IV 1-12.

    References: 1

43. Hain, G., and K. Hain

    Automatic flow chart design

    Proc. ACM 20th National Conference, 1965,
    pp. 513-523.

    References: None

44. Hallman, C. A., and R. E. Weber

    Flow chart interconnections algorithm

    IBM Technical Disclosure Bulletin, Vol. 14, No. 9,
    February 1972, pp. 2717-2720.

    References: None

45. Hardy, Robert M.

    A flow chart hardware correlation

    Digital Design Vol 3, No. 6, June 1973, pp. 46-47.

    References: None

46. Harris, Jr., C. E.

    Flowchart symbols and their usage in microfilm
    information systems

    Proc. of the 21st Annual Convention of the National
    Microfilm Association, Silver Spring, Md.,
    May 9-12, 1972, pp. II 146-147.

    References:

47. Harris, G. M.

    Flowrite: A computer generated flowchart
    technique for the NCR Century

    Computer Bulletin (GB), Vol. 12, No. 8,
    Dec. 1968, pp. 293-300.

    References: 1

48. Hawkins, R. B.

    MAFLOW - A computer language for producing
    flowcharts (Bell Telephone)

    Personal Communications with Dick B. Simmons

    References:

49. Hemdal, G.

    The function flow chart, a specification and design
    tool for SPC exchanges

    Conference on Software Engineering for Tele-
    communications and Switching Systems, Colchester,
    Essex England, April 2-5, 1973, pp. 262-270.

    References: None

50. Hirose, Ken, and Makoto Oya

Some results in general theory of flow charts

First USA-Japan Computer Conf. Proc., Oct 3-5 1972, Tokyo, Japan, pp. 12-1-1 - 12-1-5.

References: 6

51. Hirose, Ken and Makoto Oya

General theory of Flow-Charts

Comment, Math. Univ. St. Pauli XXI-2, 1972 pp. 55-71.

References: 8

52. Honeycutt, T. L.

Teaching the art of flow charting in computerized problem solving

J. Educ. Data Process Vol. 9, No. 3, 1972 pp. 19-28.

References: None

53. Houck, E. Dean, and Louis T. Copits

Flowcharts and standards (letter to editor)

Datamation, Vol. 10, No. 6, June 1964, p. 12.

References: None

54. IBM

AUTOFLOW system

Contract No. NAS5-10021 by Applied Data Research, Inc., Washington, D.C.

References:

55. IBM

System/360 flowchart

IBM Systems Reference Library File No. GH20-0199-2, 1970, 26 pp.

References: None

56. IBM

Flowcharting and block diagramming

Manual C20-8008-0 IBM Corp., 1960, 27 pp.

References:

57. IBM

Flowcharting techniques

Manual GC20-8152-1 IBM Corp., New York, March 1970, 37 pp.

References: None

58. IBM

IBM system/360 flowchart user's manual

IBM Form GH20-0293-2, IBM Corp., October 1969, 3rd Edition, New York, 56 pp.

References: None

59. IBM

Flowcharting template

Bulletin X20-8020 IBM Corp., New York, 1969 (one plastic cutout drawing guide in a printed envelope) (Update version GX20-8020-1).

References: 1

60. Jones, Robert L., and Gail Oliver

BASIC LOGIC FOR PROGRAM FLOWCHARTING AND TABLE SEARCH

Book Pub. by Anaheim Pub. Co., 1968, 88 pp.

References: None

61. Kasi, Takumi [*]

Translatability of flowcharts into while programs

Jr. of Computer and System Sciences, Vol. 9. No. 2, October 1974, pp. 177-

References:

62. Keillor, S. A.

Flow charting by interactive computer graphic

Computer Aided Design Vol. 4, No. 4, July 1972. pp. 165-168.

References: 1

63. Keller, Robert M.

A solvable program-schema equivalence problem

5th Annual Princeton Conference on Information Service & Systems 1971, pp. 301-306.

References: 5

64. King, P. J. H., and R. G. Johnson

Comments on the algorithms of Verhelst for the conversion of limited-entry decision tables to flowcharts

Comm. ACM, Vol. 17, No. 1, January 1974, pp. 43-44.

References: 6

65. Knuth, Donald E.

Computer-drawn flowcharts

Comm. ACM Vol. 6, No. 9, September 1963, pp. 555-563.

References: 7

66. Koudry, John

    Logic [flow] charting the total system.

    Data Process Magazine Vol. 8, No.1.
    January 1966, pp. 26-31.

    References: None

67. Lenher, John K.

    FLOWCHARTING: An introductory text and workbook.

    Book pub. by Auerbach, 1972, 101 pp.

    References: None

68. Lewis, F.D., and M.F.Stewart

    An automatic documentation and symbolic
    flowcharting program for the IBM S/360.

    S/360 General Program Library Prog.,
    No. 3600-00.2.003, December 1967, 86 pp.
    (Update version 00.1.014)

    References: None

69. Lewis, F. David

    Evolution of automatic flowcharting techniques.

    Unpub. IBM internal memorandum March 5,1967.

    References: None

70. Mahle,Jr., Jack D.

    Flowcharts for information retrieval system.

    National Technical Information Service AD 705 635.
    November 13, 1969, 245 pp.

    References: None

71. Maycock, P., A.C.Tory and R.N.Wintle.

    Automatic coding from a flow diagram.

    Decus Europe 8th Seminar Proceeding Strasbourg,
    France, 20-22 September 22, 1972, pp. 381-383.

    References:

72. McGee, Russell C.

    Flow charts

    Computer News Vol. 4, No. 76, May 1, 1966, pp. 1-11.

    References:

73. McQuire, Thomas V.

    APHLO automatic flowcharting program.

    Written at General Dynamics, Astronautics,
    San Diego, California.

    References:

74. McInerney, Thomas F., and Andre J.Vallee

    A STUDENT'S GUIDE TO FLOWCHARTING.

    Book pub. by Prentice-Hall Inc. 1973,
    136 pp.

    References: None

75. McQuigg, James D. and Altha M. Harness

    FLOWCHARTING.

    Book pub. by Houghton Mifflin Co., 1970, 43 pp.

    References: None

76. Michener, James C.

    The flowchart programming system.

    Simulation Vol.6,No.1, January 1971, pp.42-44

    References: 3

77. Michener, James Cope

    The flowchart programming system.

    PhD. Thesis, Brown University, June 1970, 229 pp.

    References: 25

78. Military Standard

    Flowchart symbols for automatic data
    processing systems.

    MIL-STD-682A, 21 May 1962, 11 pp.

    References: None

79. Montalabano, M.

    Tables, flowcharts and program logic.

    IBM systems, Journal Vol.1, No.1, Sept. 1962,
    pp.51-63.

    References: 3

80. Morris,D.

    The generation of programs from flowcharts.

    Proc. of the Culham Symposium on Software
    Engineering, 1971, pp. 32-37.

    References: 1

81. Morris D., T.G.Kennedy, and L.Last

    Flocoder

    The Computer Journal (British) Vol.14,No.3,
    August 1971, pp. 221-223.

    References: None

82. Nassi, J., and B.Shneiderman

Flowchart techniques for structured programming.

Sigplan Not., Vol.8, August 1973, pp. 12-26.

References: 8

83. O'Brien, F., and R.C.Beckwith

A technique for computer flowchart generation.

Computer Journal (British) Vol. 11, No.2,
August 1969, pp. 138-140.

References: 2

84. Omohundro, Dr.Wayne and Dr.James H.Tracey

FLOWWARE - A flow charting procedure to
describe digital networks.

Proc. of the First Annual Symposium on Computer
Architecrure, Edited by G.L.Lipovski and
S.A.Szygenda, Univ. of Florida, Dec. 9-11,
1973, pp. 91-97.

References: 24

85. Page, R.W.

Flow charts streamline information.

Industrial Engineering No.1
January 1969.

References:

86. Peterson, Norman D.

An "Atlas" for the visual comparison of
COBOL flowcharting software.

IEEE Repository Paper #R71-129, May 1971, 94 pp.

References: 53

87. Peterson, W.W., T.Kasami and N.Tokura

On the capabilities of while, repeat, and exit.

Comm. ACM Vol.16, No.8, August 1973, pp. 503-512.

References: 11

88. Pomeroy, Richard W.

Basic flow charting techniques.

System and Procedures Quarterly, August 1957,
pp. 2-8.

References: None

89. Repsher, William G.

Bellflow draws flow diagrams automatically (flowchart).

Bell Lab Rc. Vol.49, No.7, August 1971, pp. 209-215.

References: None

90. Robbins, M.F. and J.D.Beyer

An interactive computer system using graphical
flowchart input.

Comm. ACM. Vol.13, No.2, February 1970, pp. 115-119.

References: 31

91. Roberts, C.V.

Automated diagram documentation (ADD).

Proc. of the Fourth Meeting of UAIDE, New York,
October 1965, pp. XXIII 1-20.

References: None

92. Roberts, K.V.

The readability of computer programs.

Computer Bulletin, Vol.10, No.4, March 1967.
pp. 17-24.

References: 11

93. Rossheim, Robert J.

Report on proposed American standard flowchart
symbols for information processing.

Comm. ACM. Vol.6, No.10, October 1963, pp. 599-604.

References: None

94. Saalbach, C.P., and B.F.Sapovchak

The flowchart program.

Proc. of the Fourth Meeting of UAIDE, New York,
1965, pp. XXII 1-8.

References: 1

95. Sacks, Edward I.

Picking the best design with flowcharts.

Data Processing Magazine, Vol.8, No.12,
December 1966, pp. 22-26.

References: None

96. Schriber, Thomas J.

FUNDAMENTALS OF FLOWCHARTING.

Book pub. by John Wiley & Sons, 1969, 127 pp.

References: None

97. Scott, A.E.

Automatic preparation of flow chart listing.

Jr. ACM Vol.5, No.1, January 1958, pp. 57-66.

References: 1

98. Shelly, Gary B. and Thomas J.Cashman

    INTRODUCTION TO FLOWCHARTING AND COMPUTER
    PROGRAMMING LOGIC.

    Pub. by Anaheim Pub. Company, 1972, 244 pp.

    References: None

99. Sherman, P.M.

    FLOWTRACE, a computer program for flowcharting
    computer programs.

    Bell Telephone Labs Memo.

    References:

100. Sherman, Philip, M.

    FLOWTRACE, a computer program for flow-
    charting programs.

    Comm. ACM Vol.9, No.12, December 1966, pp. 845-854.

    References: 11

101. Simmons, D.B.

    Automated flowcharting of computer programs.

    PhD. Thesis, Univ. of Penna. 1968, 142 pp.

    References: 58

102. Sippl, Charles J.

    COMPUTER DICTIONARY and HANDBOOK.

    Book pub. by Howard W.Sams & Inc. 1966,
    Appendix N - Flowcharting - Logic, Symbols,
    Abbreviations, pp. 580-593.

    References: None

103. Smyth, M.B.

    Unique-Entry graphs, flowcharts, and state
    diagrams.

    Information and Control Vol.25, No.1,
    May 1974, pp. 20-29.

    References: 7

104. Stelwagon, W.B.

    Automatic flow charting Part 1. Using
    the program FLOW 2.

    NOTS IDP 2431 Part 1, August 1965, U.S.Naval
    Ordnance Test Station, China Lake, Calif. 22 pp.

    References: None

105. Stelwagon, W.B.

    Automatic flow charting Part 2, Principles
    and procedures.

    NOTS IDP 2431 Part 2, August 1965, U.S.Naval
    Ordnance Test Station. China Lake, Calif., 13 pp.

    References: 14

106. Stelwagon, W.B.

    Principles and procedures for the automatic
    flowcharting program FLOW 2.

    NOTS TP4095 (DOC No.AD 637 863) August 1966,
    Research Dept. U.S.Naval Ordnance Test,
    China Lake, Calif. 64 pp.

    References: 10

107. Stephens, J.V.

    Use flow charting for better test procedures.

    Power Engineering Vol.75, No.9, Sept. 1972,
    pp. 34-37.

    References: None

108. Strong, H.R.,(Jr.)

    Translating recursion equations into flow charts.

    Proc. 2nd Annual ACM Symposium on Theory of
    Computing, ACM. May 1970, pp. 184-197.

    References: 10

109. Strong, H.R.,(Jr.)

    Translating recursion equations into flow charts.

    Jr. Computer & Systems Sci. Vol.5, No.3,
    June 1971, pp. 254-285.

    References: 12

110. Strong, H.R.

    Flowchartable recursive specifications.

    Courant Computer Science Symposium 4 - Algorithm
    Specification - Edited by Randall Rustin,
    Prentice-Hall, 1972, pp. 81-96.

    References: 4

111. Swenson, Sally

    Flowchart on library search techniques.

    Special Libraries, Vol. 56, No. 4, April 1965,
    pp. 239-242.

    References: None

112. Urschler, G.

    The transformation of flow diagrams into
    maximally parallel form.

    Proc. 1973 Sagamore Computer Conf. Parallel
    Processing, Syracuse Univ,m August 22-24,
    1973, pp. 38-46.

    References: 10

113. USA Standard

    Flowchart symbols for information processing

    ASA X3.5 Approved June 8, 1966, 11 pp.

    References: 18

114. USA Standard

Flowchart symbols and their usage in
information processing.

USASI X3.5 Approved May 2, 1968, 16 pp.

References: None

115. Verhelst, M.

The conversion of limited-entry decision
tables to optimal and near-optimal
flowcharts: Two new algorithms.

Comm. ACM Vol.15, No.11, November 1972,
pp. 974-980.

References: 9

116. Waller, R.R.

Program and data flowcharts.

The Computer Bulletin (GB) Vol.15,No.10,
October 1971, pp. 372-373.

References: None

117. Watkins, R.P.

A survey of automatic flowchart generators.

Aust. Comput.J., Vol.5, No.3, November 1973,
pp. 132-140.

References: 27

118. Wayne, Mark W.

FLOWCHARTING CONCEPTS AND DATA PROCESSING
TECHNIQUES: A SELF-INSTRUCTIONAL GUIDE.

Book pub. by Canfield Press, 1973, 248 pp.

References: None

119. Weiderman, N.H., and B.M.Rawson

Flowcharting loops without loops.
Sigplan Not., Vol.10, No.4, April 1975,
pp. 37-46.

References: 5

120. Weiss, A.S.

New flowchart provides process details
at a glance.

Chem. Eng. Vol.77, No.8, April 20, 1970,
pp. 170, 172.

References: None

121. Wenker, J.

ALF-ANALGOL language flowcharts.

Datafair, Manchester, England
25-29 August 1969, 1 page abstract.

References:

122. Wettern, Herbert

The essence and advantage of standardized
programming (in German).

Electronik (Germany) Vol.21, No.9,
September 1971, pp. 317-321.

References: None

123. Yelowitz, Lawrence

Deviation of a path-connectivity
matrix for tagged flowcharts.

Jr. ACM. Vol.22, No.1, Jan. 1975, pp. 145-154

References: 11

124. Zenor, John Julian

Use of FLOW3 automatic flowcharting system to
produce flow chart from flow decks and FORTRAN
programs.

NWC TN 3030-130, May 1972, U.S.Naval Ordnance
Test Station, China Lake, Calif. 18 pp.
Appendix 40 pp.

References: None

125. ----------

Talking down a program.

Business Automation, Vol.19, No.5, May 1972,
pp. 26-27.

References: None

INDEX